Essential Clinically Applied Anatomy of the Peripheral Nervous System in the Limbs

Essential Clinically Applied Anatomy of the Peripheral Nervous System in the Limbs

Paul Rea

MBChB, MSc, PhD, MIMI, RMIP, FHEA, FRSA
University of Glasgow, Glasgow, UK

AMSTERDAM • BOSTON • HEIDELBERG • LONDON • NEW YORK
OXFORD • PARIS • SAN DIEGO • SAN FRANCISCO • SINGAPORE
SYDNEY • TOKYO
Academic Press is an imprint of Elsevier

Academic Press is an imprint of Elsevier
125, London Wall, EC2Y 5AS, UK
525 B Street, Suite 1800, San Diego, CA 92101-4495, USA
225 Wyman Street, Waltham, MA 02451, USA
The Boulevard, Langford Lane, Kidlington, Oxford OX5 1GB, UK

British Library Cataloguing-in-Publication Data
A catalogue record for this book is available from the British Library

Library of Congress Cataloging-in-Publication Data
A catalog record for this book is available from the Library of Congress

ISBN: 978-0-12-803062-2

For information on all Academic Press publications
visit our website at http://store.elsevier.com/

Working together
to grow libraries in
developing countries

www.elsevier.com • www.bookaid.org

CONTENTS

Preface...ix
Acknowledgments...xiii

Chapter 1   Overview of the Nervous System 1
1.1   Divisions of the Nervous System...1
    1.1.1   Central Nervous System 1
    1.1.2   Peripheral Nervous System 14
1.2   Functional Division of the Nervous System............................17
    1.2.1   Somatic Nervous System..17
    1.2.2   Autonomic Nervous System...18
    1.2.3   Sympathetic and Parasympathetic Nervous System 19
    1.2.4   Sympathetic Nervous System...19
    1.2.5   Parasympathetic Nervous System 21
1.3   History Taking and Clinical Examination 23
    1.3.1   Presenting Complaint 27
    1.3.2   History of Presenting Complaint 27
    1.3.3   Past Medical History 27
    1.3.4   Family History...28
    1.3.5   Social History 28
    1.3.6   Systems Review...28
1.4   Examination of the Sensory System 32
1.5   Examination of the Motor System 34
References..39

Chapter 2   Upper Limb Nerve Supply...41
2.1   Overview of the Upper Limb Nervous System 41
2.2   Branches of the Brachial Plexus 41
    2.2.1   Dorsal Scapular Nerve..45
    2.2.2   Long Thoracic Nerve..48
    2.2.3   Suprascapular Nerve 51
    2.2.4   Nerve to Subclavius 57
    2.2.5   Lateral Pectoral Nerve 61
    2.2.6   Medial Pectoral Nerve 61

2.2.7   Musculocutaneous Nerve................................63
2.2.8   Median Nerve 67
2.2.9   Medial Cutaneous Nerve of Arm...................75
2.2.10   Medial Cutaneous Nerve of Forearm 76
2.2.11   Ulnar Nerve 78
2.2.12   Upper Subscapular Nerve 84
2.2.13   Lower Subscapular Nerve 84
2.2.14   Thoracodorsal Nerve 85
2.2.15   Axillary Nerve..88
2.2.16   Radial Nerve...90
References..99

Chapter 3   Lower Limb Nerve Supply 101
3.1   Overview of the Lower Limb Nervous System 101
3.2   Cutaneous Innervation of the Lower Limb 102
       3.2.1   Subcostal Nerve 103
       3.2.2   Iliohypogastric Nerve.....................................104
       3.2.3   Ilioinguinal Nerve 105
       3.2.4   Genitofemoral Nerve 105
       3.2.5   Lateral Cutaneous Nerve of Thigh 108
       3.2.6   Cutaneous Nerve of Obturator Nerve............110
       3.2.7   Posterior Cutaneous Nerve of Thigh 111
       3.2.8   Saphenous Nerve 112
       3.2.9   Superficial Fibular Nerve 114
       3.2.10   Deep Fibular Nerve 116
       3.2.11   Sural Nerve..117
       3.2.12   Medial Plantar Nerve.....................................119
       3.2.13   Lateral Plantar Nerve.....................................119
       3.2.14   Clunial Nerves 119
3.3   Motor and Sensory Nerves of the Lower Limb.......122
       3.3.1   Superior Gluteal Nerve 122
       3.3.2   Inferior Gluteal Nerve 127
       3.3.3   Nerve to Quadratus Femoris...........................134
       3.3.4   Pudendal Nerve 135
       3.3.5   Perineal Nerve 136
       3.3.6   Inferior Rectal Nerve 137
       3.3.7   Dorsal Nerve of the Penis or Clitoris 137
       3.3.8   Nerve to Obturator Internus 137

3.3.9   Femoral Nerve 139

3.3.10   Sciatic Nerve 144

3.3.11   Obturator Nerve 151

3.3.12   Tibial Nerve 156

3.3.13   Common Peroneal Nerve 166

3.3.14   Superficial Peroneal Nerve 171

3.3.15   Deep Peroneal Nerve... 172

References.. 172

Subject Index 179

One area that has always proved to be a challenge in both undergraduate and postgraduate study of medicine, surgery, dentistry and the related health professions, and in the life sciences, has been the nervous system.

All too often, resources available in the field of anatomy, neurology and the neurosciences, for the study of the nervous system, are fraught with complex anatomical detail. On the other hand, more clinically applied resources tend to focus on the presentation of the patient rather than focusing on the underlying anatomy. Knowledge of the underpinning anatomy is absolutely essential to understand the clinical presentation in the patient. Yet, resources currently available tend to focus on one or the other, and do not combine these essential features.

Therefore, the purpose of this textbook is to provide the key anatomical facts, in an easy to access format, for each of the major regions of the nervous system related to the upper and lower limbs. Hopefully what has been achieved with this resource is a balance between anatomical content and clinical presentations, without over complicating the anatomy. It takes the reader through, in a step-by-step process, the key features of the nervous system of the upper and lower limbs.

First, the nervous system is introduced and the major divisions, both structurally and functionally, are examined. This allows the key features of the nervous system to be laid out in an approachable fashion to provide a broad scope of understanding. Then, the central and peripheral nervous systems are compared and contrasted to provide the reader with an easy to follow overview of these regions. The same applies when dealing with the somatic and autonomic nervous systems.

The following chapter then introduces the upper limb from a skeletal perspective and then provides a broad overview of the key features of the nervous system in that territory. Then, a detailed account is given of the brachial plexus, followed by the cutaneous and/or motor innervation of nerves supplying the upper limb.

Then, the lower limb is dealt with in exactly the same format to maintain continuity. This chapter introduces the lower limb from a skeletal perspective and then provides a broad overview of the key features of the nervous system in that territory. Leading on from this, the cutaneous and/or motor innervation is given in a step-by-step format for each of the main nerves in the lower limb.

What differs with this text is that rather than copious amounts of detail, the essential anatomy is presented, frequently in tables. That way the key facts related to each region or nerve can be accessed immediately without having to cross reference the text frequently.

Another key feature of this text lies in the fact that when each area is dealt with, the pertinent clinical applications are discussed. This involves detailing how to clinically examine each nerve or set of nerves. It frequently provides hints and tips on how to examine these clinically in the patient. This means that all key information is readily accessible in the same region.

Also, detailed descriptions on how to examine each of these nerves in turn are provided, with common clinical conditions encountered in clinical practice discussed. This provides direct clinical correlations to pathologies involving the anatomy of these nerves.

The combination of ensuring the relevant anatomical detail is provided next to the clinical correlations, and hints and tips on how to examine each region, makes this text incredibly unique.

The anatomical information is also supported by high-resolution digital photographs of many of the nerves being referred to throughout. A fully labeled image is provided of professionally dissected specimens (prosections) from the Laboratory of Human Anatomy within the School of Life Sciences, part of the College of Medical, Veterinary and Life Sciences in the University of Glasgow. The reader can therefore identify the anatomy of some of these important structures, and also see related anatomy in one single snapshot of that region.

It is hoped that this text will be the perfect revision tool, to have at the bedside, surgery or outpatient department, or simply to have at hand for quick access to the key anatomy, clinical applications and reminders of how to examine a patient's neurological system in relation to the upper and lower limbs for upcoming professional examinations. Pathologies of these areas frequently present to a wide variety of healthcare practitioners. It would be envisaged that this book would be a primary book as a quick reference guide, but can also be used alongside the readers' other resources from their clinical experience and training, including lecture/tutorial notes, textbooks, workbooks, etc., as it would bring together material covered in so many fields including pathology, surgery, clinical examination, and anatomy.

I really do hope you enjoy this book and find it a great companion during your studies and professional life. Good luck, whatever your reason is for using this resource.

Paul Rea
MBChB, MSc, PhD, MIMI, RMIP, FHEA, FRSA
Senior University Teacher
Laboratory of Human Anatomy
Thomson Building
School of Life Sciences
College of Medical, Veterinary and Life Sciences
University of Glasgow, Glasgow, UK

ACKNOWLEDGMENTS

There are several people whom I would like to thank in making this book possible.

First I would like to express my gratitude to Elsevier for having the time, patience, and faith in me while putting this together. They have really been the backbone to helping me realize my dream in publishing this.

I would like to dedicate this book to my mother Nancy, father Paul, and dearest brother Jaimie. Thank you for being there and supporting me throughout everything – I am so proud of you all! Thank you also to Jennifer Rea – my sister-in-law but more like a sister to me.

I would also like to extend a very special note of thanks to David Kennedy for hearing every update, word by word, for the progress of this book, and for being there in so many ways! Thank you so much!

Also, thank you to a dear friend who has gone but not been forgotten – Mark Peters.

Thank you also to Elaine Jamieson, Richard Locke, and Leana Zaccarini for all the years of a very special friendship.

Finally, thank you to a dear colleague, friend, and amazing mentor who has supported me from when I first started my career as an anatomist through to where I am today – Dr John Shaw-Dunn.

CHAPTER *1*

Overview of the Nervous System

1.1 DIVISIONS OF THE NERVOUS SYSTEM

1.1.1 Central Nervous System

The CNS is comprised of the brain and the spinal cord. The role of the CNS is to integrate all the body functions, from the information it receives. Within the peripheral nervous system (PNS), there are abundant nerves (group of many nerve fibers together), however, the CNS does not contain nerves. Within the CNS, a group of nerve fibers traveling together is called a pathway or tract. If it links the left- and right-hand sides, it is referred to as a *commissure*.

1.1.1.1 Neurons

Within the CNS, there are many, many millions of nerve cells called neurons. Neurons are cells that are electrically excitable and can transmit information from one neuron to another by chemical and electrical signals. There are three very broad classifications of neurons – sensory (which process information on light, touch, and sound to name some sensory modalities), motor (supplying muscles), and interneurons (which interconnect neurons via a network).

Typically, a neuron comprises a few simple features, though there are a variety of specializations that some have depending on the location they are found within the nervous system. In general, a neuron has a *cell body*. Here, the nucleus, or the powerhouse of the neuron, lies within its cytoplasm. At this point, numerous fine fibers enter called *dendrites*. These processes receive information from adjacent neurons keeping it up-to-date with the surrounding environment. Through these dendrites, the amount of information that a single neuron receives is significantly increased. From a neuron, there is a long single process of variable length called an *axon*. This conducts information away from the neuron, or the cell body. Some neurons however have no axons and the dendrites will conduct information to and from the neuron. In addition to this,

Essential Clinically Applied Anatomy of the Peripheral Nervous System in the Limbs
http://dx.doi.org/10.1016/B978-0-12-803062-2.00001-2

a *lipoprotein* layer called the *myelin sheath* can surround the axon of a principal cell. This is not a continuous layer along the full length of the axon. Rather, there are interruptions called *nodes of Ranvier*. It is at this point where the voltage-gated channels occur, and it is here that conduction occurs. Therefore, the purpose of the myelin sheath is to enable almost immediate conduction between one node of Ranvier and the next, thus ensuring quick communication between neurons, and keeping those nerves *up to date* with body processes around them.

The size of neurons however varies considerably. The smallest of our neurons can be as small as 5 μm, and the largest, for example, motor neurons, can be as large as 135 μm. In addition, axonal length can vary considerably too. The shortest of these can be 100 μm, whereas a motor axon supplying the lower limb, for example, the toes, can be as long as 1 m.

In the PNS, neurons are found in *ganglia*, or in *laminae* (layers) or *nuclei* in the central nervous system (CNS).

Neurons communicate with each other at a point called a *synapse*. Most of these junctional points are chemical synapses where there is the release of a *neurotransmitter* that diffuses across the space between the two neurons. The other type of synapse is called an *electrical synapse*. This form is generally more common in the invertebrates, where there is close apposition of one cell membrane and the next, that is, at the pre- and postsynaptic membranes. Linking these two cells is a collection of tubules called *connexons*. The transmission of impulses occurs in both directions and rapidly. This is because there is no delay in the neurotransmitter having to be activated and released across the synapse. Instead, the flow of communication depends on the membrane potentials of the adjacent cells.

1.1.1.2 Neuroglia

Neuroglia, or glia, are the supportive cells for neurons. Their main purpose is not in relation to the transmission of nerve impulses. Rather, they are involved in provision of nutrients, maintenance of a stable homeostatic environment, and the production of the myelin sheath. There are two broad classifications of neuroglia – *microglia* and *macroglia*.

Microglia have a defensive role and are known as *phagocytic cells*. They are found throughout the brain and spinal cord, and can alter their shape, especially when they engulf particulate material. Therefore, they function in a protective role for the nervous system.

Macroglia are subdivided into seven different types, again with each having a special role.

1. *Astrocytes*: These cells fill the spaces between neurons and provide for structural integrity. They also have processes that join to the capillary blood vessels. These are known as perivascular end feet. Therefore, with their close apposition to the vasculature, they are also thought to be responsible for metabolite exchange between the neurons and the blood vessels. They are found in the CNS.
2. *Ependymal cells*: There are three types of ependymal cells – ependymocytes, tanycytes, and choroidal epithelial cells. The ependymocytes allow for the free movement of molecules between the cerebrospinal fluid (CSF) and the neurons. Tanycytes are generally found in the third ventricle of the brain and can be involved in responding to changing hormonal levels of the blood-derived hormones in the CSF. Choroidal epithelial cells are the cells that control the chemical composition of the CSF. They are found in the CNS.
3. *Oligodendrocytes*: These cells are responsible for the production of myelin sheaths. They are found in the CNS.
4. *Schwann cells*: Like oligodendrocytes, Schwann cells are responsible for the production of the myelin sheath, but in the PNS. They also have an additional role in phagocytosis of any debris; therefore help to clean the surrounding environment.
5. *Satellite cells*: These cells surround those neurons of the autonomic system and also the sensory system. They maintain a stable chemical balance of the surrounding environment to the neurons. Therefore they are found in the PNS.
6. *Radial glia*: Radial glial cells act as a means of scaffolding onto which new neurons migrate. They are found in the CNS.
7. *Enteric glia*: These cells are found within the gastrointestinal tract and aid digestion and maintenance of homeostasis. They are, by their very nature, found in the PNS.

In the CNS, several different types of neuron are found. The most common type of neuron found in the CNS are called *multipolar neurons*, because of having many types of dendrites, as well as the single axon. The summary of the main types of neurons is as follows.

- *Multipolar neurons*: These have at least two dendrites that extend from the neuronal soma. Multipolar neurons are also classified as *Golgi Type 1 neurons*, and *Golgi Type 2 neurons*. Golgi Type 1 neurons have long axons that originate in the gray matter of the spinal cord. Golgi Type 1 neurons are typically found in the ventral gray horn of the spinal cord. They are also typical of *pyramidal neurons* of the cerebral cortex or *Purkinje* neurons of the cerebellum. On the other hand, Golgi Type 2 neurons either do not have an axon at all, or if they do, the axon does not exit from the gray matter of the CNS. Golgi Type 2 neurons are found in the granular layer of the cerebellum and hippocampus. These neurons are called *granule cells*, typically found throughout the cerebral cortex, cerebellum, hippocampus, olfactory bulb, and the dorsal cochlear nucleus.
- *Bipolar neurons*: These neurons are less typical and have an axon and dendrite (or extension of the axon) at opposite sides of the neuronal cell body. These neurons are typically involved in transmission of information related to the special senses. They are found in the *retina* (transmission of visual information), *olfactory epithelium* (transmission of information related to smell), and the *vestibulocochlear nerve* (transmitting information related to sound and balance).
- *Unipolar neurons*: These are also referred to as *pseudounipolar* neurons. They have a single axon that extends both centrally and peripherally. The central portion of this neuron extends into the spinal cord and the peripheral portion will extend into the periphery, terminating perhaps in the skin, muscle, or joints. *Pseudounipolar* neurons do not have dendrites and are typically found in the *dorsal root ganglia*.
- *Anaxonic neurons*: These are neurons where no obvious axon is identifiable from the dendritic tree and are typically found within the retina and the brain.
- *Betz neurons*: These neurons are the largest of all neurons and are found within the primary motor cortex (of the frontal lobe). These neurons send their axons through the corticopsinal tract, to reach the ventral horn cells.

Another classification that exists for neurons is interneurons. These are neurons that connect two neurons together and can either be motor or sensory. They can also be classified as excitatory or inhibitory (more common in the CNS). Interneurons can also be thought of as local circuit neurons. Examples of interneurons are given below.

- *Spindle cells*: They are defined on the basis of their appearance. They have also been referred to as Von Economo neurons. Their soma is spindle shaped in appearance and has a single axon at the apex of the cell with a single dendrite running in the opposite direction. Spindle cells are found in the fronto-insular cortex and anterior cingulate cortex. More recently, they have also been found in the dorsolateral prefrontal cortex (Fajardo et al., 2008).
- *Lugaro cells*: They are the inhibitory sensory interneurons located within the cerebellum. These neurons interconnect a large number of neurons within the cerebellum.
- *Basket cells*: These neurons are inhibitory interneurons located within the cerebral cortex as well as the hippocampus and the cerebellum. These neurons use the inhibotry transmitter GABA as their way to communicate with other neurons. They have also been subclassified into three different subtypes – large, small, and nest basket cells on the basis of their appearance (Wang et al., 2002).
- *Unipolar brush cells*: These neurons have only one dendrite but at the end point of this, many short *brush-like* structures arise from it, thus increasing the surface area for them to communicate with. These neurons are excitatory and use the excitatory transmitter glutamate. They are typically found within the cerebellar cortex (granular layer).

It must also be noted that in the PNS, neurons are found in ganglia that is a collection of nerve cell bodies. These are found either in layers, or laminae, or in groups called nuclei in the CNS.

1.1.1.3 Gray and White Matter

In the CNS, there are two clear differences between the structural components. It is divided into its appearance of either gray or white matter. Within the gray matter, there are cell bodies and dendrites of efferent neurons, glial cells (supportive), fibers of afferent neurons, and interneurons. The white matter, on the other hand, primarily consists of myelinated axons and the supportive glial cells. The purpose of the white matter is to allow communication from one part of the cerebrum to the other, and also to communicate to other brain areas and carry impulses through the spinal cord.

1.1.1.4 Brain
The brain is a mass of convoluted neural tissue and is referred to as the *cerebrum* (Latin: brain). The brain is a complex organ consuming approximately 15% of cardiac output, and can only survive a few minutes deprived of oxygen. If it is deprived of oxygen, death will ensue quickly. It is comprised of two cerebral hemispheres – left and right, and together these arise from the embryologic *telencephalon*. The cerebral hemispheres process information related to a wide variety of functions, and will be dealt with at subsequent points later in the text. Between these two massive cerebral hemispheres lies the *diencephalon* which consists of the *thalamus* and *hypothalamus*. The telencephalon and the diencephalon together form the forebrain – the first part of the brain.

The middle part of the brain, the *midbrain*, or *mesencephalon*, structurally comprises the *tectum, tegmentum, cerebral aqueduct, cerebral peduncles*, and many nuclei and pathways (or fasciculi). It typically deals with *alertness*, the *sleep/wake cycle, hearing, vision, motor function*, and some *homeostatic regulations* like temperature control internally.

The last part of the brain is referred to as the *hindbrain* and is comprised structurally of two individual components – the *metencephalon* (pons, cerebellum, and some cranial nerve nuclei) and the mylencephalon (medulla oblongata). Clinically, it is easier to refer to the brainstem that really is the last part of the brain and is composed of the *midbrain + pons + myelencephalon* (medulla oblongata). In older texts, the diencephalon is included, but for our purposes, the aforementioned description will suffice.

The brain can also be summarized by the following subdivisions:

1. *Telencephalon* (cerebral hemispheres) + *Diencephalon* (thalamus and hypothalamus) = *Forebrain*
2. *Mesencephalon* = *Midbrain*
3. *Metencephalon* (pons, cerebellum and the trigeminal, abducent, facial and vestibulocochlear nerves) + *Myelencephalon* (medulla oblongata) = *Hindbrain*

The brainstem is directly connected with the spinal cord at the *foramen magnum*. This site is crucially important as it acts as a conduit for information passing between the periphery and the central processing unit of the brain. If there is an increase in pressure within the cranial cavity, for example, due to a *space-occupying lesion (SOC)*, *intracranial hemorrhage* or *traumatic brain injury*, it can result in pressure at the foramen magnum. The only way the brain has to move in this enclosed space is downward through the foramen magnum. This is called *brain herniation*, or *coning*, results in pressure on the brainstem, and affects the components in that part of the brain, that is, cardiorespiratory functions. This is a life-threatening condition and requires urgent neurosurgical attention.

The brain is comprised, like the spinal cord, of two types of matter – gray and white. The gray matter of the brain is found in the outer aspect of the brain in the cerebral cortex. It is comprised of neurons, whereas the white matter is composed of supporting glial cells and myelinated axons. The white appearance (especially in tissue that has been fixed in formaldehyde) is due to the lipids of the myelin. The main substance of the brain is white matter but dispersed throughout it are areas of gray matter referred to as the basal ganglia. These subcortical nuclei are designed for a variety of functions including voluntary motor control, emotions, cognition, eye movements, and learning.

1.1.1.5 Forebrain

Surrounding the core of the forebrain, that is, the diencephalon, comprises of the two large cerebral hemispheres (left and right) that constitute the cerebrum. The cerebrum is composed of three regions:

1. *Cerebral cortex*: The cerebral cortex is the gray matter of the cerebrum. It is comprised of three parts based on its functions – motor, sensory, and association areas. The motor area is present in both cerebral cortices. Each one controls the opposite side of the body, that is, the left motor area controls the right side of the body, and vice versa. There are two broad regions – a primary motor area responsible for execution of voluntary movements, and supplementary areas involved in selection of voluntary movements.

   The sensory area receives information from the opposite side of the body, that is, the right cerebral cortex receives sensory information from the left side of the body. In essence, it deals with auditory information (via the primary auditory cortex), visual information (via the primary visual cortex), and sensory information (via the primary somatosensory cortex).

The association areas allow us to understand the external environment. All of the cerebral cortex is subdivided into lobes of the brain. These are:
a. *Frontal lobes*: Broadly speaking, the frontal lobe deals with *executive* functions and our long-term memory. It also is the site of our primary motor cortex, toward its posterior part.
b. *Parietal lobes*: The parietal lobes are responsible for integration of sensory functions. It is the site of our primary somatosensory cortex.
c. *Temporal lobes*: The temporal lobes integrate information related to hearing, and therefore, are the sites of our primary auditory cortex.
d. *Occipital lobes*: The occipital lobes integrate our visual information and functions as the primary visual cortex.
2. *Basal ganglia*: The basal ganglia are three sets of nuclei – the *globus pallidus*, *striatum*, and *subthalamic nucleus*. These nuclei are found at the lower end of the forebrain and are responsible for voluntary movement, development of our habits, eye movements, and our emotional and cognitive functions.
3. *Limbic system*: The limbic system is comprised of a variety of structures on either side of the thalamus. It serves a variety of functions including long-term memory, processing of the special sense of smell (olfaction), behavior, and our emotions.

1.1.1.6 Thalamus
The thalamus is like a junction point of information. It is a relay point for all sensory information (apart from that related to smell). It also functions in the regulation of our wakened state, or sleep. In addition, it provides a connection point for motor information on its way to the cerebellum.

1.1.1.7 Hypothalamus
The hypothalamus, as its name suggests, is located below the thalamus. It secretes hormones influencing the pituitary gland, and in turn, a wide

variety of bodily functions. It regulates autonomic activity ranging from temperature control, hunger, our circadian rhythm, and thirst.

1.1.1.8 Midbrain

The midbrain, as its name suggests, is found between the lower part of hindbrain and the upper part of the cerebral cortices. Comprised of the *cerebral peduncles*, *cerebral aqueduct*, and the *tegmentum*, it is involved in motor function, arousal state, temperature control, and visual and hearing pathways.

1.1.1.9 Hindbrain

Developmentally, the lowest part of the brain is the hindbrain and comprises the pons, medulla, and the cerebellum. These areas control movement, cardiorespiratory functions, and a variety of bodily functions like hearing and balance, facial movement, swallowing, and bladder control. Therefore, brainstem death, that is, death of these regions, is incompatible with life.

1.1.1.10 Spinal Cord

The spinal cord is a long cylinder that occupies the upper two-thirds of the *vertebral canal*. Unlike the brain, the spinal cord gray matter is located within the main substance of it, and is surrounded by the white matter. It is the opposite way round in the brain. On the lateral aspects of the spinal cord, is a pair of *spinal roots*. Each side is composed of a ventral and dorsal root depending on whether it arises from the anterior (ventral) or posterior (dorsal) aspect of the spinal cord.

In summary, there are 31 pairs of spinal roots with their corresponding dorsal and ventral roots. There are eight *cervical*, twelve *thoracic*, five *lumbar*, five *sacral*, and one *coccygeal*. Each of these combinations of dorsal and ventral roots joins to form a single spinal nerve that further divides into a *dorsal* and *ventral ramus*. The specific details of the spinal cord will be explained further.

The gray matter is surrounded by the white matter that contains primarily the axons of myelinated interneurons. These groups of axons, or pathways, run longitudinally up and down to and from the brain, or between upper and lower levels of the spinal cord.

Surrounding the spinal cord are three layers of protective meninges referred to as *pia mater, arachnoid mater*, and *dura mater*. The *dural layer* is the outermost layer, and extends from approximately the second sacral segment all the way to the foramen magnum, and surrounds the brain too. The arachnoid layer is deeper, and is tightly adherent to the dura mater. It leaves a space just beneath it called the subarachnoid space. It is in this space that the fluid circulates around the brain and spinal cord passes to cushion, support, protect, and provide nourishment to the brain and spinal cord – the CSF. The innermost layer is called the pia mater and is tightly adhered to the spinal cord, and also the brain.

The spinal cord extends from the brain to the second lumbar segment, but the spinal nerves still emerge from the corresponding vertebrae all the way to the coccyx. Toward the lower end of the spinal cord, there is an enlargement called the lumbosacral segment that constitutes the *cauda equina* (Latin for horse's tail) and is that part of the lower spinal cord that extends from the second to fifth lumbar vertebral nerves. The spinal cord terminates as the *filum terminale*. This point where the spinal cord terminates has a clinical relevance – *lumbar puncture*.

As the core of the spinal cord terminates at approximately the second vertebral level, the clinician is able to insert a needle below this point to take a sample of CSF. At the third and fourth vertebral level space, a needle can be inserted into the subarachnoid space to take a sample of CSF. This can be taken for analysis to diagnose a potential infection of the meninges – *meningitis* can be potentially life threatening.

When the spinal cord is examined in transverse section, it is composed of a central gray matter (butterfly shaped) comprising cell columns oriented along the rostrocaudal axis (containing neuronal cell bodies, dendrites, and axons that are both myelinated and unmyelinated), surrounded by the white matter comprising the ascending and descending myelinated and unmyelinated fasciculi (tracts). The general layout of the spinal cord is shown in Figure 1.1.

In each half of the spinal cord there are three funiculi: the dorsal funiculus (between the dorsal horn and the dorsal median septum), the lateral funiculus (located where the dorsal roots enter and the ventral roots exit), and the ventral funiculus (found between the ventral median fissure and the exit point of the ventral roots).

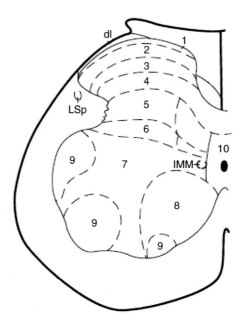

Fig. 1.1. Cross-section of the right side of the spinal cord indicating the position of Rexed's laminae. 1–10, the positions of laminae I–X, respectively; dl, dorsolateral funiculus; IMM, intermediomedial nucleus; LSp, lateral spinal nucleus.

Based on detailed studies of neuronal soma size (revealed using the Nissl stain), Rexed (1952) proposed that the spinal gray matter is arranged in the dorsoventral axis into laminae and designated them into 10 groupings of neurons identified as I–X.

Lamina I contains the terminals of fine myelinated and unmyelinated dorsal root fibers that pass first through the zone of Lissauer (dorsolateral funiculus) and then enter lamina I mediating pain and temperature sensation (Christensen and Perl, 1970; Menétrey et al., 1977; Craig and Kniffki, 1985; Bester et al., 2000). The neurons here have been divided into small neurons and large marginal cells characterized by wide-ranging horizontal dendrites (Willis and Coggeshall, 1991). They then synapse on the posteromarginal nucleus. From here, the axons of these cells pass to the opposite side and ascend as the lateral spinothalamic tract.

Lamina II is found just beneath lamina I, and is referred to as the *substantia gelatinosa*. Neurons here modulate the activity of pain and temperature afferent fibers, though intrinsic neurons here do not contain the target for substance P, the NK-1 receptor (Bleazard et al., 1994) that is found in lamina I, III, and IV (Naim et al., 1997). Lamina II has been subdivided into an outer (dorsal) lamina II (II_O) and an inner (ventral) lamina II (II_i) based on the morphology of these layers. Stalked cells were found in larger numbers in lamina II_O but stalked and islet cells were found throughout lamina II (Todd and Lewis, 1986). Indeed, lamina II_i was also found to be different in its neurochemical profile. Lamina II is the region that receives an extensive unmyelinated primary afferent input, with very little from large myelinated primary afferents (except for the distal parts of hair follicle afferents in some animals; Willis and Coggeshall, 1991). The axonal projections from here are wide and varied with some neurons projecting from the spinal cord (projection neurons), some going to different laminae, and some with axons confined to a lamina in the region of the dendritic tree of that cell, for example, intralaminar interneurons, local interneurons, and Golgi Type II cells (Todd, 1996).

Lamina III is distinguished from lamina II in that it has slightly larger cells, but with a neuropil similar to that of lamina II. The classical input to this lamina comes from hair follicles and other types of coarse primary afferent fibers that include Pacinian corpuscles and rapidly and slowly adapted fibers.

Lamina IV is a relatively thick layer that extends across the dorsal horn. Its medial border is the white matter of the dorsal column, and its lateral border is the ventral bend of laminae I – III. The neurons in this layer are of various sizes ranging from small to large and the afferent input here is from collaterals and from large primary afferent fibers (Willis and Coggeshall, 1991). Input also arises from the substantia gelatinosa (lamina II) and contributes to pain, temperature and crude touch via the spinothalamic tract (Siegel and Sapru, 2006).

Lamina V extends as a thick band across the narrowest part of the dorsal horn. It occupies the zone often called the neck of the dorsal horn. It has a well-demarcated edge against the dorsal funiculus, but an indistinct lateral boundary against the white matter due to the many longitudinally oriented myelinated fibers coursing through this area. The cell types are very homogeneous in this area, with some being slightly larger than in lamina IV (Willis and Coggeshall, 1991). Again, like lamina IV, primary afferent input into this region is from large primary afferent collaterals as well as receiving descending fibers from the corticospinal and rubrospinal tracts with axons also contributing to the spinothalamic tracts (Siegel and Sapru, 2006). In addition, in the thoracolumbar segments (T1–L2/3), the reticulated division of lamina V contains projections to sympathetic preganglionic neurons (Cabot et al., 1994).

Lamina VI is present only in the *cervical* and *lumbar* segments. Its medial segment receives joint and muscle spindle afferents, with the lateral segment receiving the rubrospinal and corticospinal pathways. The neurons here are involved in the integration of somatic motor processes.

Lamina VII is found in the intermediate region of the spinal gray matter and contains Clarke's nucleus extending from C8 to L2. This nucleus receives tendon and muscle afferents with the axons of Clarke's nucleus forming the dorsal spinocerebellar tract relaying information to the ipsilateral cerebellum (Snyder et al., 1978). Also within lamina VII are the sympathetic preganglionic neurons constituting the intermediolateral cell column in the thoracolumbar (T1–L2/3) and the parasympathetic neurons located in the lateral aspect of the sacral cord (S2–S4). In addition, Renshaw cells are located in lamina VII and are inhibitory interneurons that synapse on the alpha motor neurons and receive excitatory collaterals from the same neurons (Renshaw, 1946; Siegel and Sapru, 2006).

Lamina VIII and IX are found in the ventral gray matter of the spinal cord. Neurons here receive descending motor tracts from the cerebral cortex and the brainstem, and have both alpha and gamma motor neurons here that innervate skeletal muscles (Afifi and Bergman, 2005). Somatotopic organization is present where those neurons innervating the extensor muscles are ventral to those innervating the flexors, and neurons innervating the axial musculature are medial to those innervating muscles in the distal extremities (Siegel and Sapru, 2006).

Lamina X is the gray matter surrounding the central canal and represents an important region for the convergence of somatic and visceral primary afferent input conveying nociceptive and mechanoreceptive information (Nahin et al., 1983; Honda, 1985; Honda and Perl, 1985). In addition, lamina X in the lumbar region also contains preganglionic autonomic neurons as well as an important spinothalamic pathway (Ju et al., 1987a,b; Nicholas et al., 1999).

Table 1.1 provides an easy to follow summary of the input, output, and information processed by the 10 laminae found within the spinal cord.

1.1.2 Peripheral Nervous System

The PNS is the part of the nervous system that is comprised of the *cranial, spinal, and peripheral nerves,* as well as their sensory and motor

Table 1.1 Input of Each One of Rexed's Laminae, the Destination of Neurons in Each Layer, and an Overview of the Functions of Each of the Territories in the Spinal Cord

	Input	Destination	Information Processed
I	Fine myelinated and unmyelinated dorsal root fibers	Lateral spinothalamic tract	Pain and temperature sensation
II	Unmyelinated primary afferent input	Projection neurons Variety of laminae Confined to the laminae of the dendritic tree of the neuron	Modulate the activity of pain and temperature afferent fibers
III	Hair follicles Pacinian corpuscles Rapidly and slowly adapted fibers	Deeper spinal laminae Posterior column nuclei Supraspinal relay centers	Mechanoreception Propriospinal pathways Pain, temperature and touch
IV	Collaterals Large primary afferent fibers Substantia gelatinosa	Spinothalamic tract	Pain, temperature and crude touch
V	Large primary afferent Collaterals Corticospinal and rubrospinal tracts	Spinothalamic tract	Pain and temperature sensation
VI	Descending corticospinal and rubrospinal fibers (lateral segment) Joint and muscle spindle afferents (medial segment)	Innervation of limbs	Integration of somatic motor processes
VII	Tendon and muscle afferents	Spinocerebellar tract Sympathetic ganglia (in thoracic and upper lumbar regions) Parasympathetic fibers in S2–S4	Proprioception Visceral (autonomic) regulation
VIII	Descending motor tracts from the cerebral cortex and the brainstem	Motor neurons	Intrafusal muscle fibers
IX	Descending motor tracts from the cerebral cortex and the brainstem	Skeletal muscles	Posture and balance Distal muscle movement
X	Convergence of somatic and visceral primary afferent input Autonomic regulation (in lumbar region)	Nociceptive and mechanoreceptive information	Nociceptive and mechanoreceptive information

nerve endings. In other words, it is the part of the nervous system that is comprised of nerves and ganglia that are present within the brain and spinal cord (which comprises the CNS). As this part of the nervous system is primarily present within the skull and vertebral column, it is prone to damage from trauma and from toxins.

Therefore, the PNS is comprised of the 12 pairs of cranial nerves, and the 31 pairs of spinal nerves, that is, 43 pairs of nerves in total. The nerves of the PNS can be classified as belonging to either afferent (taking information to the CNS) or efferent (away from the CNS). Spinal nerves contain both afferent and efferent information, whereas some cranial nerves like the olfactory and optic nerves contain only afferent information (for smell and sight, respectively).

Broadly speaking, there are two main divisions of the PNS – the somatic and the autonomic nervous systems (ANS). The somatic nervous system terminates on the skeletal muscle, whereas the ANS supplies all structures other than the skeletal muscle, for example, glands and smooth muscle.

1. Efferent
   a. *Special visceral efferent*: Motor fibers that target skeletal muscle. These fibers target muscles of pharyngeal origin. They are specifically carried in the trigeminal, facial, glossopharyngeal, vagus, and accessory cranial nerves.
   b. *Somatic efferent*: These are motor fibers that target skeletal muscle.
   c. *General visceral efferent*: These fibers are motor fibers that are autonomic in origin, for example, cardiac, glandular, and smooth muscle.
2. Afferent
   a. *General somatic afferent*: These fibers are sensory fibers that carry information related to general sensation, for example, touch, pain, and temperature.
   b. *General visceral afferent*: These fibers carry sensory information related to the visceral organs. They arise from the glands, blood vessels, and the viscera specifically. They are generally classified as those belonging to the ANS. The cranial nerves that carry these impulses are the facial, glossopharyngeal, and vagus nerves.

c. *Special visceral afferent*: These fibers carry information related to pharyngeal arch origin related to the gastrointestinal tract. They carry information related to taste and smell. These fibers are carried in cranial nerves, specifically the olfactory, facial, glossopharyngeal, and vagus nerves.

1.1.2.1 Spinal Nerves

There are 31 pairs of spinal roots with their corresponding dorsal and ventral roots. There are eight *cervical*, twelve *thoracic*, five *lumbar*, five *sacral*, and one *coccygeal*. Within the spinal roots, there are dorsal and ventral roots. In the ventral roots, there are motor fibers, and it is those fibers that supply the skeletal muscle. Within the ventral roots of the thoracic, upper lumbar, and some of the sacral levels autonomic fibers are also present. Within the dorsal roots, there are sensory fibers from the skin, subcutaneous, and deep tissue, and frequently from the viscera too. The spinal nerve is formed by both the dorsal and ventral root and contains most of the fiber components found in that root. Therefore, a major peripheral nerve contains the sensory, motor, and autonomic fibers within it. However, smaller branches will vary in their composition. Therefore, a nerve to skin will lack motor fibers to the skeletal muscle, but it will contain sensory fibers and autonomic fibers to the vasculature and may also contain fibers supplying the autonomic innervation to the hair follicles.

The dorsal rami will transmit information from the muscles of the back and also from the skin. The ventral rami innervate the rest of the trunk and also the limbs. The ventral rami that supply the thorax and abdomen remain relatively separate in their course. However, in the cervical or lumbar and sacral regions, the ventral rami are found intertwined in what is called a plexus of nerves. It is from these plexuses that the peripheral nerves will then emerge.

A simple way to look at it is that when the ventral rami enter into this plexus, its individual components will contribute to several of the peripheral nerves. Therefore, each peripheral nerve will indeed contain fibers from one or more spinal nerves.

Each spinal nerve has a distribution called a *dermatome*. This is the area that is supplied by a single spinal nerve by its sensory fibers running in its dorsal root. However, sectioning of a single spinal nerve rarely will result in complete loss of sensation, or anesthesia. This is because other adjacent spinal nerves will also be carrying fibers from that site. Therefore, the more likely scenario is a reduced level of sensation, or hypoesthesia.

There are many versions of *dermatome maps* that can be used clinically to demonstrate the site(s) of pathology of a patient with a suspected spinal-nerve lesion. The most common one to be used clinically is the American Spinal Injury Association's (ASIA) worksheet produced as the International Standards for Neurological Classification of Spinal Cord Injury (ISNCSCI). This is discussed in considerable detail in Chapter 2.

Motor fibers present in the ventral root tend to supply more than one muscle. Therefore, each muscle will receive innervation from more than one spinal nerve. Sectioning of a single spinal nerve will result in weakness of more than one muscle. Sectioning of a peripheral nerve will however result in paralysis of that single muscle. The group of muscles that a single spinal nerve supplies is called a *myotome*. Testing of the myotomes is something routinely carried out in the neurological examination of a patient, directed according to the signs and symptoms that the person presents.

1.2 FUNCTIONAL DIVISION OF THE NERVOUS SYSTEM

Figure 1.2 summarizes the functional divisions of the nervous system. Central to this diagram is the brain and spinal cord (CNS). Information from the periphery arriving into the CNS is referred to as afferent. Information exiting the CNS is referred to as efferent. There are two afferent inputs that enter the CNS – somatic and visceral.

TIP!

The easy way to remember what information arrives into and leaves the CNS is rather easy.
*A*FFERENT – *A*RRIVES into the CNS. *A* for *AFFERENT*, *A* for *ARRIVES!*
*E*FFERENT – *E*XITS the CNS. *E* for *EFFERENT*, *E* for *EXIT!*

1.2.1 Somatic Nervous System
The somatic nervous system consists of the cell bodies located in either the brainstem or the spinal cord. They have an extremely long course

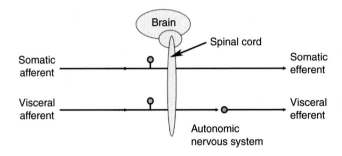

Fig. 1.2. *This diagram shows the functional divisions of the nervous system into its somatic and visceral components. Information arriving into the CNS is referred to as afferent, and information leaving it is efferent. Note the 2-neuron pathway in the autonomic (visceral) nervous system.*

as they do not synapse after they leave the CNS until they are at their termination in skeletal muscle. They consist of large diameter fibers and are ensheathed with myelin. They are commonly referred to as motor neurons because of their termination in skeletal muscle. Within the muscle fibers, they release the neurotransmitter acetylcholine and are only excitatory, that is, result only in contraction of the muscle.

1.2.2 Autonomic Nervous System

The visceral, or ANS, can be thought of as that part of the nervous system supplying all other structures apart from skeletal muscle (supplied by the somatic nervous system). However, part of the ANS supplies the gastrointestinal system and is referred to as the enteric nervous system as the neurons are found supplying the glands and smooth muscle in the actual wall of the tract.

Within the ANS generally, it is composed of two neurons and a synapse (as shown in Figure 1.2). This is different to the single neuron of the somatic nervous system. The origin of the first neuron of the ANS is found in the CNS, with the first synapse occurring in an *autonomic ganglion*. This part is defined as the *preganglionic fiber*. After the synapse in the autonomic ganglion, the second fiber is referred to as the *postganglionic fiber* as it passes to the *effector* organ, in this case *cardiac* or *smooth muscle*, *glands*, or *gastrointestinal neurons*.

The ANS is subdivided into *sympathetic* and *parasympathetic* divisions on the basis of physiological and anatomical differences. The *sympathetic* division arises from the *thoracolumbar* region from the first thoracic to the second lumbar level (T1–L2). The *parasympathetic* division arises from *cranial* and *sacral* origins. Specifically, the parasympathetic division arises from four of the cranial nerves – the *oculomotor* (III), *facial* (VII), *glossopharyngeal* (IX), and *vagus* (X) nerves. It also arises from the *sacral plexus* at the levels of the second to fourth sacral segments (S2–S4).

1.2.3 Sympathetic and Parasympathetic Nervous System

The sympathetic nervous system is part of the nervous system that deals with *fight or flight* responses. The parasympathetic nervous system can be classified as part of the nervous system that controls *rest and digest*. A summary table is given in the subsequent section comparing what functions each part of the ANS causes to a variety of areas around the body (Table 1.2).

1.2.4 Sympathetic Nervous System

The sympathetic nervous system arises from the thoracolumbar region of the spinal cord. Most of the sympathetic ganglia are present in close proximity to the spinal cord forming two chains on either side of the body. These are referred to as the *sympathetic trunks*. However, some ganglia are present a little further away from the spinal cord and are referred to as *collateral ganglia*. These are found close to the arteries in the abdomen with the same names, that is, coeliac, superior mesenteric, and inferior mesenteric ganglia. They tend to be closer to the organs that they supply.

Although the sympathetic nervous system arises specifically at the thoracolumbar region, that is, from the first thoracic to either the second or third lumbar vertebral levels, the sympathetic trunk extends from the neck to the sacrum. This is because some of the preganglionic fibers arising from the thoracolumbar region travel either up or down several vertebral segments before forming their synapses with the respective postganglionic neurons. Within the neck, the cervical ganglia are referred to as the superior and middle cervical or stellate ganglia.

This not only allows the sympathetic nervous system to act as a single unit but also enables small areas to act independently. This contrasts the parasympathetic nervous system that tends to act independently. This arrangement is ideal to be involved in fine regulation of the activities of the organs or territories that they supply.

Table 1.2 Comparison and Contrast of the Wide Effects of the Sympathetic and Parasympathetic Nervous System at Key Areas Throughout the Body

	Sympathetic	Parasympathetic
Heart	Raising of heart rate	Reduction of heart rate
	Increase of contractility of the atria and ventricles	Reduction in contractility of the atria and ventricles
	Increased conduction	Reduced conduction
Lungs	Relaxation of the bronchial muscles	Contraction of the bronchial muscles
	Reduction in secretions (via $\alpha 1$ receptors)	Stimulation of secretions (via $\alpha 1$ receptors)
Stomach and intestines	Reduction in tone and motility	Increase of tone and motility
	Contraction of sphincters	Relaxation of sphincters
	Inhibition of secretions	Stimulation of secretions
Pancreas	Inhibition of exocrine secretion	Stimulation of exocrine secretion
	Inhibition of insulin secretion	Stimulation of insulin secretion
Eyes	Contraction of radial muscle (dilatation of the pupil)	Contraction of the sphincter muscle (constriction of the pupil)
	Relaxation of the ciliary muscle (for far vision)	Contraction of the ciliary muscle (for near vision)
Nasal, lacrimal and salivary glands	No significant effect	Stimulation of mucous and serous secretions from the secretory cells
Skin	Contraction of the arrector pili muscles (hair to stand on end)	N/A
	Localized secretion of the sweat glands	Generalized secretion of the sweat glands
Urinary bladder	Relaxation of the bladder wall	Contraction of the bladder wall
	Contraction of the sphincter	Relaxation of the sphincter
Genital organs	May stimulate vasoconstriction, but it is uncertain and variable	May stimulate glands and smooth muscle; vascular dilatation
Adrenal gland	Stimulation of secretory cells to produce epinephrine	No effect
Arterioles	Variable	Dilation of the coronary and salivary gland arterioles (via $\alpha 1,2$ receptors)

The sympathetic nervous system is responsible for the body's *fight or flight* reaction. Therefore, through its innervation of the adrenal medulla that releases adrenaline (epinephrine) as its major secretion (80%; with the other 20% being noradrenaline (norepinephrine)), it helps to protect the body in times of threat to it. Therefore, it would be involved in functions like dilating the pupil, increasing heart rate and contractility, relaxation of the bronchial muscle, reduction in secretion of the bronchial glands, and reduction of gut motility. This allows blood to be diverted to those areas in need if the body needs to *fight or flight*. The adrenal medulla is a bit unusual in its innervation by the sympathetic nervous system, as the postganglionic side of the adrenal medulla never develops axons. Instead the preganglionic fibers terminating in the adrenal medulla results in the secretion from it of epinephrine/norepinephrine and is viewed as an endocrine gland as its secretions pass into the bloodstream.

1.2.5 Parasympathetic Nervous System

The parasympathetic nervous system is described as originating in the craniosacral region, that is, from the brainstem and also from the sacral plexus. Specifically, the parasympathetic nervous system cranially is concerned with three of the cranial nerves that will be dealt with in turn in greater detail throughout this book. The cranial nerves involved in the parasympathetic nervous system are the oculomotor, facial, glossopharyngeal, and vagus nerves. Specifically, the nuclei related to these are the Edinger–Westphal nucleus for the oculomotor nerve, superior salivatory and lacrimal nuclei for the facial nerve, inferior salivatory nucleus for the glossopharyngeal nerve, the dorsal nucleus of the vagus nerve, and the nucleus ambiguus for the vagus nerve. This is where the preganglionic fibers are found for the parasympathetic nervous system. In addition to this, the sacral parasympathetic nucleus arising from the second, third, and fourth sacral segments are also involved.

The parasympathetic nervous system is opposite in its functions generally to the sympathetic nervous system. It can informally be referred to as the part of the nervous system responsible for *rest and digest*, that is, responsible for the internal functions when you are sitting resting and relaxing. Therefore, it would constrict the pupil, slow heart rate and contractility, contract bronchial musculature and stimulate bronchial secretions, and enhance gut motility for digestion to effectively occur.

The main neurotransmitter in both the sympathetic and parasympathetic nervous systems at the preganglionic fiber, as it contacts the postganglionic fiber, is acetylcholine. The same is also true at the postganglionic fiber as it contacts the effector organ generally. Therefore, where acetylcholine is secreted, it is referred to as *cholinergic*. However, in the sympathetic nervous system, the major neurotransmitter between the postganglionic fiber and the effector organ tends to be noradrenaline (norepinephrine). It also tends to be the case that this is not an exclusive relationship as to what is secreted, and at what site it is secreted. In addition to this, cotransmitters tend to also be present, for example, ATP, dopamine, and other neuropeptides.

Table 1.2 summarizes the differences between the sympathetic and parasympathetic nervous systems at a variety of regions throughout the body.

1.2.5.1 Cranial Nerves

There are 12 pairs of cranial nerves that are associated with the brain. These nerves are unique in the human body and may carry one, some, or many different types of fibers within them. Cranial nerves are very different to spinal nerves (or roots) in that they do not contain dorsal and ventral roots. Indeed, some cranial nerves may have one ganglion, several, or none at all. Indeed, the optic nerve, responsible for conveying the sensation for vision, is deemed a direct outgrowth of the brain, and can be compared to a fiber tract from the CNS.

Cranial nerves arise from the brain, as distinct from spinal nerves that arise from the spinal cord. There are 12 pairs of cranial nerves. Some are purely motor (e.g., hypoglossal (XII) that supply the tongue muscles); some are purely sensory (e.g., the optic (II) nerves that come from the retina); some are mixed (e.g., the trigeminal (V) nerve that is sensory to the face and scalp and motor to the muscles of mastication). We can consider the following:

1. *Where they are attached to the brain:*
   a. *Forebrain:* Olfactory (I) and optic (II) nerves
   b. *Midbrain:* Oculomotor (III) and trochlear (IV) nerves. (The trochlear is unique in arising from the dorsal surface of the brain stem.)
   c. *Hindbrain:* Trigeminal (V) from the pons; abducens (VI), facial (VII), and vestibulocochlear (VIII) from the pontomedullary junction; glossopharyngeal (IX), vagus (X), accessory (XI), and hypoglossal (XII) from the medulla.
2. *How they are related to embryological development:* Many of the structures in the head and neck arise from two quite distinct embryological sources. These sources are
   a. *Somites:* These are paired segmental blocks of tissue that run along the length of the embryo, rather like a series of building

Table 1.3 The Cranial Nerves Associated with Each of the Branchial Arches, and the Muscles They Supply

Arch		Nerve	Muscles
I	Mandibular arch	Trigeminal (V)	Mastication
II	Hyoid arch	Facial (VII)	Facial expression
III–VI		IX, X, XI	Laryngeal/pharyngeal

bricks. They give rise to many structures, including muscles. Motor cranial nerves that supply somite-derived muscles are called somatic efferent* nerves and consist of cranial nerves III, IV, VI that supply the extraocular muscles (which move the eyes about) and cranial nerve XII to the muscles of the tongue.

b. *Branchial arches*: During development, the embryo passes through a stage of having pharyngeal or branchial arches at the side of the neck – exactly as a fish has gill arches. These arches form a numbered series and, again, give rise to many adult structures, including muscles. Motor cranial nerves that supply them are termed branchial efferent nerves. They are given in Table 1.3.

1.3 HISTORY TAKING AND CLINICAL EXAMINATION

The following should be recorded when undertaking a medical history of the patient:

1. Name
2. Date of birth
3. Gender
4. Occupation
5. Source of the history – typically the patient but perhaps from another individual if the patient is unconscious. If another individual is giving the history of the patient, the involvement of this person should be recorded, for example, member of the public, professional, witness of incident, paramedic, etc.

---

*"Efferent" means "going away from." In the context of cranial or spinal nerves, efferent nerves are those going out from the CNS, that is, motor nerves. The opposite is "afferent" (means going toward) that would describe sensory nerves taking information to the CNS. Table 1.4 summarizes the cranial nerves in what they supply, and the pathways.

Table 1.4 Key Features of Each of the Twelve Pairs of Cranial Nerves

Nerve	Components	Functions	Point of Entry/ Exit from Brain	Exits/Enters Cranial Cavity	Nuclei	Ganglion	Important Branches
Olfactory (I)	Special sensory	Smell	Forebrain	Cribriform plate of ethmoid bone	No specific nucleus. Olfactory epithelium contain the cell bodies	None	Olfactory epithelium (central processes)
Optic (II)	Special sensory	Vision	Midbrain	Optic canal	Lateral geniculate nucleus	Retinal ganglion cells	Optic nerve; optic tract
Oculomotor (III)	Somatic motor Visceral motor	Extra-ocular muscles Sphincter muscle and ciliary muscle	Midbrain	Superior orbital fissure	Oculomotor nucleus; Edinger–Westphal nucleus	Ciliary ganglion	Motor branches to extra-ocular muscles; parasympathetic division
Trochlear (IV)	Somatic motor	Innervates the superior oblique muscle	Midbrain	Superior orbital fissure	Nucleus of the trochlear nerve	None	None. Only supplies the superior oblique muscle
Trigeminal (V)	General sensory; branchial motor	Sensation from face; paranasal sinuses; nose and teeth Muscles of mastication	Pons	Superior orbital fissure (Va), foramen rotundum (Vb) or foramen ovale (Vc)	Spinal trigeminal nucleus; pontine trigeminal nucleus; mesencephalic trigeminal nucleus; trigeminal motor nucleus	Trigeminal ganglion; Submandibular ganglion	Ophthalmic nerve; maxillary nerve; mandibular nerve
Abducent (VI)	Somatic motor	Innervates the lateral rectus muscle	Pontomedullary junction	Superior orbital fissure	Abducent nerve nucleus	None	None. Only supplies the lateral rectus muscle

Nerve	Components	Functions	Location	Exit	Nuclei	Ganglia	Branches
Facial (VII)	Branchial motor; Visceral motor; Special sensory; General sensory	Muscles of facial expression, stylohyoid, stapedius, posterior belly of digastric; Parasympathetic innervation of the submandibular and sublingual salivary glands, lacrimal gland and the nasal and palatal glands; Anterior two-thirds of the tongue (and palate); Concha of the auricle	Pontomedullary junction	Stylomastoid foramen	Facial motor nucleus; lacrimal nucleus; superior salivatory nucleus; gustatory nucleus; spinal trigeminal nucleus	Geniculate ganglion; pterygopalatine ganglion; submandibular ganglion	Intratemporal Greater petrosal nerve; nerve to stapedius; chorda tympani Extratemporal Temporal; zygomatic; buccal; marginal mandibular; cervical; posterior auricular; posterior belly of digastric branch; stylohyoid branch
Vestibulocochlear (VIII)	Special sensory	Balance for the vestibular component; hearing for the spiral (cochlear) component	Pontomedullary junction	Internal auditory meatus	Vestibular nucleus; ventral cochlear nucleus; dorsal cochlear nucleus; superior olivary nucleus	Vestibular ganglion; spiral ganglion	Vestibular nerve; cochlear nerve
Glossopharyngeal (IX)	Branchial motor; Visceral motor; Special sensory; General sensory; Visceral sensory	Stylopharyngeus; Parotid gland for parasympathetic innervation; Taste from the posterior one-third of the tongue; External ear; Pharynx; parotid gland; middle ear; carotid sinus and body	Medulla oblongata	Jugular foramen	Nucleus ambiguus; solitary nucleus; spinal trigeminal nucleus; inferior salivatory nucleus	Inferior ganglion; otic ganglion; superior ganglion; inferior ganglion	Muscular; tympanic; pharyngeal; tonsillar; carotid sinus branch

(Continued)

Table 1.4 Key Features of Each of the Twelve Pairs of Cranial Nerves (*cont.*)

Nerve	Components	Functions	Point of Entry/ Exit from Brain	Exits/Enters Cranial Cavity	Nuclei	Ganglion	Important Branches
Vagus (X)	Branchial motor; Visceral motor; Special sensory; General sensory; Visceral sensory	Pharyngeal constrictors; laryngeal muscles (intrinsic); palatal muscles; upper two-thirds of oesophagus; Heart; trachea and bronchi; gastrointestianl tract; Taste from the palate and the epiglottis; Auricle; external auditory meatus; posterior cranial fossa dura mater; Gastrointestinal tract (to last one-third of the transverse colon); pharynx and larynx; trachea and bronchi; heart	Medulla oblongata	Jugular foramen	Dorsal nucleus of the vagus nerve; nucleus ambiguus; solitary nucleus; spinal trigeminal nucleus	Superior ganglion; inferior ganglion	Meningeal branch; auricular branch; pharyngeal branches; superior laryngeal nerve; recurrent laryngeal nerve; cardiac branches; cardiac branches; esophageal branches; pulmonary branches; gastric branches; celiac branches; renal branches
Spinal Accessory (XI)	Somatic motor	Innervates the sternocleidomastoid and trapezius muscles	Medulla oblongata (and spinal cord)	Jugular foramen	Nucleus ambiguus; spinal accessory nucleus	None	Cranial branch; Spinal branch
Hypoglossal (XII)	Somatic motor	Extrinsic and intrinsic muscles of the tongue. Palatoglossus is not supplied by the hypoglossal nerve. It is supplied by the glossopharyngeal nerve	Medulla oblongata	Hypoglossal canal	Hypoglossal nucleus	None. It may however receive general sensory fibers from the ganglion of C2	Meningeal branches; thyrohyoid branches; muscular branches

1.3.1 Presenting Complaint

Typically, this can be written down in the patient's own words and will be the reason for presenting to the medical or allied healthcare practitioner.

1.3.2 History of Presenting Complaint

The following information should be included:

1. What are the symptoms/signs?
2. Which region of the body is affected?
3. When did the signs/symptoms commence?
4. What is the duration of signs/symptoms?
5. Was the onset gradual or sudden?
6. Is there a history of trauma?
7. Have there been any exacerbating or relieving factors?
8. Are there any other associated features?

Specific to a neurological examination, any other related features should also be noted, for example:

1. Headache
2. Nausea and/or vomiting
3. Numbness/tingling
4. Weakness, loss of balance, or stiffness of joints/limbs
5. Visual disturbance
6. Altered smell
7. Any episodes of dizziness or loss of balance
8. Reduced or altered consciousness

1.3.3 Past Medical History

This can be broadly divided into the following areas:

1. *Medical*: It is important to record any newly diagnosed diseases, or ones that the patient may have had in the past, or for some time. This can include, but not limited to, hypertension, diabetes mellitus, asthma, hay fever, previous myocardial infarction, angina, epilepsy, etc.
2. *Surgical*: The procedures undertaken and their dates, any complications that have/had arisen, and why they were performed should be recorded.
3. *Obstetric/gynecological/sexual health*: This should include a full obstetric history and menstrual history as relevant, as well as any birth controls used currently or previously. A sexual health can be

taken if it is indicated to the presentation (e.g., number of sexual partners, type of sex, risky activities).
4. *Psychiatric*: This should include the type of illness and its duration, any periods of hospitalization, diagnosis, and treatments provided.

1.3.4 Family History

It may be relevant to draw a family tree if indicated in the history taking. It should include the relation to the individual of the condition presenting to relatives. It should also detail the age of the individual(s) and general health, and if there is a notable family history, and how it has affected the relatives of the patient.

Again, this should include any history of hypercholesterolemia, hypertension, diabetes mellitus, asthma, hay fever, previous myocardial infarction, angina, epilepsy, genetic conditions, etc.

1.3.5 Social History

For this section of the medical history, it should include the following:

1. *Alcohol consumed*: How much, when it is taken during the day, if it is consumed on the patient's own or with others around?
2. *Smoking*: How much and for how long has the patient smoked?
3. Any use of illicit drugs, and if so what was consumed and for how long the person has adopted this choice of drug in his/her lifestyle?
4. *Occupation*: This should involve analysis of any occupational risk factors like exposure to toxins or other chemical agents
5. Relationship status to identify the home background of the patient. It may be relevant to ask about the sexual orientation of the patient and sexual practices.
6. Exercise and dietary factors should be asked about, as well as any use of alternative therapies.

1.3.6 Systems Review
1.3.6.1 Cardiovascular (CVS)

The following should be included when recording the CVS history of the patient:

1. Any type of *heart complaint* or issues with their heart
2. Hypertension

3. Dyspnea
4. Chest pain or discomfort
5. Edema
6. Heart murmurs
7. Palpitations
8. Paroxysmal nocturnal dyspnea
9. Any history of cardiac investigations, when they occurred, and also the results of such investigations

1.3.6.2 Respiratory (RS)

The following should be included when recording the CVS history of the patient:

1. Cough
2. Sputum
3. Wheezing
4. Dyspnea
5. Hemoptysis
6. Bronchitis
7. Emphysema
8. Pleurisy
9. Any recent or historical chest X-rays, their results, as well as any other relevant imaging or diagnostic interventions undertaken

1.3.6.3 Gastrointestinal System (GI)

The following should be included when recording the CVS history of the patient:

1. Nausea
2. Vomiting
3. Dysphagia
4. Weight loss
5. Altered bowel habits
6. Hematemesis
7. Abdominal pain or discomfort and the exacerbating factors (or relieving ones)
8. Excessive flatulence or gas formation
9. Jaundice

10. Liver or gallbladder problems
11. Hemorrhoids
12. Rectal bleeding

1.3.6.4 Head, Eyes, Ears, Nose and Throat (HEENT)

The following should be included when recording the CVS history of the patient:

Head

1. Headaches
2. Dizziness
3. Trauma to the head
4. Loss of consciousness
5. Altered vision
6. Pain

Eyes

1. Visual defects, for example, double vision, loss of sight in one or both eyes (temporary or permanent)
2. Use of corrective glasses/contact lenses
3. Pain or red eye
4. Glaucoma, cataracts, or any other eye condition

Ears

1. Hearing and any loss of, or increased noise
2. Vertigo
3. Tinnitus
4. Earache
5. Discharge

Nose

1. Difficult in breathing
2. Patency
3. Discharge
4. Stuffiness
5. Itching
6. Bleeding

Throat

1. Pain
2. Discharge
3. Swelling
4. Teeth and their condition
5. Dentures and any issues related to implants

1.3.6.5 Neurological

The following should be included when recording the CVS history of the patient:

1. Numbness
2. Loss or reduction in power of muscles
3. Paresthesia (pins and needles)
4. Weakness, or noticeable contractures or difficulty moving a muscle or muscle group
5. Altered sensation
6. Paralysis of an area or areas of the body
7. Fainting
8. Blackouts or seizures
9. Mood
10. Attention
11. Memory
12. Changes in speech
13. Headaches

1.3.6.6 Genitourinary

The following should be included when recording the CVS history of the patient:

1. Frequency of urination
2. Dysuria
3. Hematuria
4. Nocturia
5. Polyuria
6. Urgency
7. Retention
8. Hesitancy

9. Dribbling
10. Incontinence
11. Menstrual cycle – duration, onset of menstruation/cessation
12. Lumps and bumps, for example, groin or genitalia
13. Discharge
14. Number of children
15. Sexual intercourse and any related issues, for example, pain, infections etc.

1.3.6.7 Musculoskeletal (MSK)

The following should be included when recording the CVS history of the patient:

1. Arthritis
2. Difficulty moving joints
3. Muscle or joint pain
4. Pain, tenderness, red joints, and if so then where
5. Reduced range of motion of joints
6. Backache
7. Onset of any symptoms and their duration and any exacerbating or relieving factors

1.4 EXAMINATION OF THE SENSORY SYSTEM

In terms of examining the nervous system of a patient, there are several key areas that should be noted.

1. *Observation of the patient*: Here the following should be noted when examining the patient:
   a. *Gait*: The patient should be observed either while walking into the room, or while they are by the bedside. Make sure to examine any imbalance, footdrop, ataxia, or the shuffling gait of Parkinson's disease. Also observe any noticeable muscle contractures or asymmetry.
   b. *Speech*: When listening to the patient, listen to the tone of voice, observe facial symmetry and muscle activity, and be observant for sentence construction, hoarseness, or absence of the voice. Note any dysarthria, aphasia, or dysphasia.
2. *Sensory examination*: Both the upper and lower limbs can be examined as indicated by the patient's presenting complaint and the

history of the conditions described by the patient. These should be relevant to what is being investigated but it should also be noted that nothing should be omitted if there is clinical suspicion or uncertainty about the patient.

The sensory systems of the limbs can be assessed using the International Standards for Neurological Classification of Spinal Cord Injury (ISNCSCI). This enables assessment and classification of both motor and sensory function and is classified as follows:

Sensory assessment

Light touch and pinprick are assessed separately, and a score out of two is given for each dermatome on the right and left side of the body.

The dermatomes that are assessed are cervical (C2, C3, C4, C5, C6, C7, C8), thoracic (T1, T2, T3, T4, T5, T6, T7, T8, T9, T10, T11, T12), lumbar (L1, L2, L3, L4, L5), and sacral (S1, S2, S3, S4–S5). It is imperative to ensure that this is completed comprehensively as indicated by the patient's complaint.

For sensation, the following scoring system is used as highlighted in Table 1.5.

A score is given for light touch right (LTR totaling 56, i.e., 2 for each of the vertebral levels stated previously) and light touch left (LTL totaling 56, i.e., 2 for each of the vertebral levels stated previously). This is recorded as follows:

$$LTR + LTL = \text{out of } 112$$

A score is also given for pinprick right (PPR totaling 56, i.e., 2 for each of the vertebral levels stated previously) and pinprick left (PPL

Table 1.5 Socring System Used for Sensation Testing	
Sensory score	Classification
0	Absent
1	Altered
2	Normal
NT	Not testable

totaling 56, i.e., 2 for each of the vertebral levels stated previously). This is recorded as follows:

$$PPR + PPL = \text{out of } 112$$

In addition to light touch and pinprick, temperature sensation should be assessed as well as proprioception and vibration sense of each territory.

Temperature can be simply assessed in the clinical environment with a tuning fork that is generally cooler on the skin, although it may be necessary to test hot and cold sensation using water of differing temperatures.

Proprioception can typically be assessed at the distal interphalangeal joints of the hand and feet, for example, index finger and great toe. Vibration sense may be tested, though is not always conclusive.

1.5 EXAMINATION OF THE MOTOR SYSTEM

The following broad outline can be used to assess the motor system of both upper and lower limbs, functioning of the motor system, and the related musculature in each region. Specific details will be given in subsequent chapters.

1. *Inspection*: When examining the motor system on inspection of the patient, make sure you observe the following:
   a. Posture
   b. Muscle atrophy or hypertrophy
   c. Muscle fasciculation
2. *Tone*: Typically, hypotonia is found with a lower motor neuron lesion and also with disease of the cerebellum. Hypertonia is found with upper motor neuron lesion pathologies.
   In addition a lower motor neuron lesion will also have muscle fasciculation, reduced reflexes, and perhaps muscle paralysis. With an upper motor neuron lesion, there may be muscular weakness and spasticity, and also a positive Babinski sign. This means rather than plantar flexion on stroking the foot with a blunt object, the reverse happens and there will be plantar extension. This is an indication of disease in the brain/spinal cord and is present as a primitive reflex in infants, but disappears after approximately 1 year.

3. *Power*: The power of muscles should be tested against resistance to ensure that an accurate examination is undertaken. Again, like the sensory examination, the motor examination is covered in the International Standards for Neurological Classification of Spinal Cord Injury (ISNCSCI), and is described as follows:

Each major muscle group is assessed and examined in turn of the vertebral levels.

For the upper limbs, the left and right sides should be assessed, compared, and contrasted:

a. C5 – Elbow flexors
b. C6 – Wrist extensors
c. C7 – Elbow extensors
d. C8 – Finger flexors
e. T1 – Finger abductors

For the lower limbs, the left and right sides should be assessed, compared, and contrasted:

a. L2 – Hip flexors
b. L3 – Knee extensors
c. L4 – Ankle dorsiflexors
d. L5 – Long toe extensors
e. S1 – Ankle plantar flexors

For each muscle group, on the left and right hand sides, a score of 5 is given for each on the basis of its function using the following classification in Table 1.6.

Table 1.6 The Motor Score as well as the Findings on Clinical Examination of Each Muscle Group in Turn

Motor score	Classification of muscle function
0	Total paralysis
1	Palpable or visible contraction
2	Active movement, with gravity eliminated
3	Active movement against gravity
4	Active movement, against some resistance
5	Active movement, against full resistance
5*	Normal but corrected for pain/disuse
NT	Not testable

Therefore while assessing a patient without pathology of the musculature and its innervation, the following formula would result in 50/50 for the upper extremities and 50/50 for the lower extremities in total. This is because a mark out of 5 would be given for each upper-limb side (right/left) for C5 (elbow flexors), C6 (wrist extensors), C7 (elbow extensors), C8 (finger flexors), and T1 (finger abductors). This results in total score of 25 for the upper-limbs (right) side and 25 for the upper-limbs (left) side. In addition, a mark out of 5 would be given for each lower-limb side (right/left) for L2 (hip flexors), L3 (knee extensors), L4 (ankle dorsiflexors), L5 (long toe extensors), and S1 (ankle plantar flexors), that is,

Upper extremities right (UER) 25 (out of a maximum of 25)

+

Upper extremities left (UEL) 25 (out of a maximum of 25)

=

50

Lower extremities right (UER) 25 (out of a maximum of 25)

+

Lower extremities left (UEL) 25 (out of a maximum of 25)

=

50

4. *Tendon reflexes*: There are two types of reflexes – deep and superficial tendon jerk reflexes. When recording the tendon reflexes in clinical case notes, it should be recorded as follows:

+++ Hyperactive reflex

++ Normal reflex

+ Sluggish reflex

− Absent reflex

Deep tendon jerk reflex

a. *Upper limbs*: In relation to the upper limbs, the following should be examined for deep tendon jerk reflexes:

- *Biceps tendon jerk reflex (C5/6)*: With the patient's arm relaxed, the examiner should place their thumb and remaining fingers over the biceps tendon and the tendon hammer should then lightly tap the thumb of the examiner, as it lies over the biceps tendon.

- *Triceps tendon jerk reflex (C6/7)*: The patient should move their arm over their chest and relax it. The examiner should then tap the tendon of the triceps tendon over the elbow joint to elicit this response.
- *Supinator tendon jerk reflex (C5/6)*: Just superior to the wrist joint, and with the patient lightly placing their hand over their abdomen, the examiner should lightly tap the area of the supinator muscle to elicit this response.

b. *Lower limbs*: In relation to the lower limbs, the following should be examined for superficial tendon jerk reflexes:
- *Knee jerk reflex (L3/4, and occasionally L2)*: The patient should have their knee flexed, and the examiner can place their hand into the popliteal fossa.
- *Ankle jerk (S1)*: The patient should be lying down on the examining couch/bed with the knee in flexion and the examiner should dorsiflex the ankle joint and with the leg slightly laterally rotated. The examiner should then tap the tendon hammer over the Achilles tendon to elicit the reflex.
- Superficial tendon jerk reflex:
  Lower limbs: In relation to the lower limbs, the following should be examined for superficial tendon jerk reflexes:
  Plantar response: With the patient lying flat, a blunt object can be rolled across the lateral aspect of the foot starting at the heel and then moving toward the great toe. A normal response will be plantar flexion. As described earlier, an extensor plantar response would be a positive Babinski sign, and is not seen unless the patient is less than 1 year old. A Babinski sign would indicate an upper motor neuron lesion.

TIP!

If it proves difficult to elicit the ankle jerk reflex from the Achilles tendon, the patient can be asked to kneel on a chair, facing the back of it. The legs should be dangling from the chair, and tapping over the Achilles tendon can then help elicit the reflex.

5. *Coordination*

a. Upper limbs
   - *Finger–nose test*: The examiner should place their index finger approximately 20–30 cm away from the eye level of the patient. The patient should then be instructed to touch their nose and then the examiner's finger. Then, the patient should be instructed to move their own index finger between their nose and the examiner's finger. To increase the level of difficulty of this test, the examiner can then move their index finger, and ask the patient to touch their nose and the moving index finger of the examiner.
   - To test for sensory ataxia, ask the patient to close their eyes and then touch the tip of their nose with an outstretched finger.
   - In both examinations, observe for intention tremor and also past-pointing. Past-pointing means that when the patient approaches the examiner's index finger with their own, they miss the examiner's digit. These features can suggest cerebellar disease.
   - *Check for dysdiadochokinesis*: Ask the patient to touch one dorsal surface of the hand with the palmar surface of the opposite hand. The opposite hand should then rotate to the dorsal surface of the opposite hand. This alternating palmar/dorsal surface onto the opposite hand should be repeated as rapidly as possible for the patient. Dysdiadochokinesis is the inability to undertake this rapid movement and can indicate cerebellar disease.

b. Lower limbs
   - *Assess gait*: Ask the patient to walk from one side of the room (or examining area) to the other. If they normally use an aid to walking, they should be allowed to do so.
   - *Heel to toe*: The patient should be asked to walk forward by placing one heel in front of the toes then switching to the opposite side and to keep walking in this fashion for a short distance.
   - *Romberg's test*

      i. Ask the patient to stand up with their feet together, arms by their side and eyes open.

ii. Then, ask the patient to close their eyes for approximately 20–30 s.

iii. The patient may exhibit mild swaying which is normal.

iv. It is possible to repeat the test two times to help assessment. If the patient loses their balance, it is said that they have a positive Romberg's test, or Romberg's sign.

Therefore, these clinical examinations can be applied to both the upper limbs and lower limbs to assess the integrity of the neuromuscular functioning of each of these regions. Specific details for each nerve will be given, as relevant, in each of the subsequent chapters and subchapters. Clinical examination of the cranial nerves has not been dealt with, since the purpose of this text is to deal with the essential clinically applied anatomy of the PNS in the limbs. Further details of the cranial nerves, clinical examination, and associated pathologies can be found in the companion text in this series – Clinically Applied Anatomy of the Cranial Nerves (Rea, 2014).

REFERENCES

Afifi, A., Bergman, R., 2005. Functional Neuroanatomy, second ed. McGraw-Hill, USA.

Bester, H., Chapman, V., Besson, J.M., Bernard, J.F., 2000. Physiological properties of the lamina I spinoparabrachial neurons in the rat. J. Neurophysiol. 83, 2239–2259.

Bleazard, L., Hill, R.G., Morris, R., 1994. The correlation between the distribution of the NK-1 receptor and the action of tachykinin agonists in the dorsal horn of the rat indicates that substance P does not have a functional role on substantia gelatinosa (lamina II) neurons. J. Neurosci. 14 (12), 7655–7664.

Cabot, J.B., Alessi, V., Carroll, J., Ligorio, M., 1994. Spinal cord lamina V and lamina VII interneuronal projections to sympathetic preganglionic neurons. J. Comp. Neurol. 347, 515–530.

Christensen, B.N., Perl, E.R., 1970. Spinal neurons specifically excited by noxious or thermal stimuli: marginal zone of the dorsal horn. J. Neurophysiol. 33, 293–307.

Craig, A.D., Kniffki, K.D., 1985. Spinothalamic lumbosacral lamina I cells responsive to skin and muscle stimulation in the cat. J. Physiol. 365, 197–221.

Fajardo, C., Escobar, M.I., Buriticá, E., Arteaga, A., Umbarila, J., Casanova, M.F., Pimienta, H., 2008. Von Economo neurons are present in the dorsolateral (dysgranular) prefrontal cortex of humans. Neurosci Lett. 435 (3), 215–218.

Honda, C.N., 1985. Visceral and somatic efferent convergence onto neurons near the central canal in the sacral spinal cord of the cat. J. Neurophysiol. 53, 1059–1078.

Honda, C.N., Perl, E.R., 1985. Functional and morphological features of neurons in the midline region of the caudal spinal cord of the cat. Brain Res. 340, 285–295.

Ju, G., Hökfelt, T., Brodin, E., Fahrenkrug, J., Fischer, J.A., Frey, P., Elde, R.P., Brown, J.C., 1987a. Primary sensory neurons of the rat showing calcitonin gene-related peptide immunoreactivity and their relation to substance P, somatostatin, galanin, vasoactive intestinal polypeptide and cholecystokinin immunoreactive ganglion cells. Cell Tissue Res. 247, 417–431.

Ju, G., Melander, T., Ceccatelli, S., Hökfelt, T., Frey, P., 1987b. Immunohistochemical evidence for a spinothalamic pathway co-containing cholecystokinin and galanin like immunoreactivities in the rat. Neurosci. 20, 439–456.

Menétrey, D., Giesler, Jr., G.J., Besson, J.M., 1977. An analysis of response properties of spinal dorsal horn neurons to non-noxious and noxious stimuli in the rat. Exp. Brain Res. 27, 15–33.

Nahin, R.L., Madsen, A.M., Giesler, G.J., 1983. Anatomical and physiological studies of the grey matter surrounding the spinal cord central canal. J. Comp. Neurol. 220, 321–335.

Naim, M., Spike, R.C., Watt, C., Shehab, A.S., Todd, A.J., 1997. Cells in laminae III and IV of the rat spinal cord that possess the neurokinin-1 receptor and have dorsally directed dendrites receive a major synaptic input from tachykinin containing primary afferents. J. Neurosci. 17 (14), 5536–5548.

Nicholas, A.P., Zhang, X., Hökfelt, T., 1999. An immunohistochemical investigation of the opioid cell column in lamina X of the male rat lumbosacral spinal cord. Neurosci. Lett. 270, 9–12.

Rea, P., 2014. Clinical anatomy of the cranial nerves, first ed. Academic Press, Elsevier, San Diego, USA.

Renshaw, B., 1946. Central effects of centripetal impulses in axons of spinal ventral roots. J. Neurophysiol. 9, 191–204.

Rexed, B., 1952. The cytoarchitectonic organisation of the spinal cord in the cat. J. Comp. Neurol. 96, 414–495.

Siegel, A., Sapru, H.N., 2006. Essential Neuroscience. Lippincott Williams and Wilkins, Baltimore.

Snyder, R.L., Faull, R.L., Mehler, W.R., 1978. A comparative study of the neurons of origin of the spinocerebellar afferents in the rat, cat and squirrel monkey based on the retrograde transport of horseradish peroxidase. J. Comp. Neurol. 15, 833–852.

Todd, A.J., 1996. GABA and glycine in synaptic glomeruli of the rat spinal dorsal horn. Eur. J. Neurosci. 8, 2492–2498.

Todd, A.J., Lewis, S.G., 1986. The morphology of Golgi-stained neurons in lamina II of the rat spinal cord. J. Anat. 149, 113–119.

Wang, Y., Gupta, A., Toledo-Rodriguez, M., Wu, C.Z., Markram, H., 2002. Anatomical, physiological, molecular and circuit properties of nest basket cells in the developing somatosensory cortex. Cereb Cortex. 12 (4), 395–410.

Willis, W.D., Coggeshall, R.E., 1991. Sensory mechanisms of the spinal cord, second ed. Plenum Press, New York.

CHAPTER 2

Upper Limb Nerve Supply

2.1 OVERVIEW OF THE UPPER LIMB NERVOUS SYSTEM

The upper limb, like the lower limb, is connected by a girdle to the trunk, and has three segments – arm, forearm, and hand. The shoulder girdle, formed by the clavicles and the scapulae, is completed anteriorly by the manubrium of the sternum, with which the medial sides of the two clavicles articulate and is deficient behind. The upper limb is characterized by a great deal of mobility, although perhaps the ball and socket joint formed between the humerus and the scapula is not the most stable of joints in the body. Many of the movements of the upper limb depend on the support and the stability provided by muscles that have an extensive origin from ribs and vertebrae. Therefore the muscles of the pectoral region and the superficial muscles of the back are also included in the description of the upper limb.

The upper limbs are first seen in the embryo at approximately 5 mm in length, that is, 4 weeks postovulation. Each limb bud elongates and develops in the proximal to distal sequence, that is, the forearm appears before the hand does. A few days later, the nerves then start to grow into the limb buds. Then, the skeleton and the muscles become differentiated. Shortly after this point, the fingers can then be recognized. The bones of each region of the upper limb are highlighted in Table 2.1.

2.2 BRANCHES OF THE BRACHIAL PLEXUS

The nerves to the upper limb arise from the brachial plexus that is a large and very important structure located partly in the neck and partly in the axilla. The brachial plexus is formed by the union of the ventral rami of the lower cervical nerves (i.e., C5, C6, C7, C8) and the greater part of the ventral ramus of the first thoracic nerve (i.e., T1). Therefore, it can be written that the origin of the brachial plexus arises from C5 to T1. However, the brachial plexus can frequently receive contribution

Essential Clinically Applied Anatomy of the Peripheral Nervous System in the Limbs
http://dx.doi.org/10.1016/B978-0-12-803062-2.00002-4

Table 2.1 The Bones Within Each of the Territories of the Upper Limb

Region of Upper Limb	Bones
Shoulder girdle	Scapula Clavicle
Shoulder joint	• *Glenohumeral joint*: Head of humerus and the glenoid cavity of the scapula • *Acromioclavicular joint*: Clavicle and the acromion of the scapula • *Sternoclavicular joint*: Clavicle and the sternum • *Scapulothoracic joint*: Scapula and ribs (2–7). Physiological joint. • *Suprahumeral joint*: Head of humerus and the coracoacromial ligament. Physiological joint.
Arm	Humerus
Elbow joint	Distal humerus Proximal radius and ulna
Forearm	Radius Ulna
Wrist joint	Distal radius Proximal carpal bones, apart from the pisiform, that is, triquetrum, lunate, and scaphoid

Hand	Carpal bones (8)
	• Proximal – scaphoid
	  – Lunate
	  – Pisiform
	  – Triquetrum
	• Distal – trapezium
	  – Trapezoid
	  – Capitate
	  – Hamate
	Metacarpals (5)
	Phalanges (14)

from the fourth cervical nerve (C4) above, or the second thoracic nerve (T2) below.

When the fourth cervical nerve is large and the first thoracic nerve is small, the plexus is described as being *prefixed*. If there is more contribution from the first and second thoracic nerves, the brachial plexus is described as being postfixed. If the first rib is rudimentary, the second thoracic nerve can provide the brachial plexus with a larger innervation.

Following from its origin, the brachial plexus then descends into the lower part of the neck called the posterior triangle. The posterior triangle of the neck has the following boundaries:

1. *Anterior*: The posterior border of the sternocleidomastoid muscle
2. *Posterior*: The anterior border of the trapezius muscle

3. *Base*: The middle third of the clavicle
4. *Apex*: The sternocleidomastoid and the trapezius, at the superior nuchal line of the occipital bone
5. *Roof*: The superficial layer from the deep cervical fascia

The brachial plexus is found above the clavicle, and posterior and lateral to the sternocleidomastoid muscle. It is posterosuperior to the third part of the subclavian artery, and is crossed by the lower belly of the omohyoid muscle. In terms of surface anatomy, the brachial plexus can be felt in both superior and inferior to the omohyoid, located in the angle between the clavicle and the sternocleidomastoid. From the surface anatomy perspective, the brachial plexus can be found in the neck below a line drawn from the posterior margin of the sternocleidomastoid at the level of the cricoid cartilage to the midpoint of the clavicle.

Superficial to the brachial plexus within the neck are several structures, namely the supraclavicular nerves, platysma, external jugular vein, inferior belly of omohyoid and the descending scapular, and transverse cervical arteries.

The brachial plexus then passes posterior to the medial two-thirds of the clavicle, running with the axillary artery as it passes deep to the pectoralis major. The cords of the brachial plexus are arranged around the axillary artery posterior to the pectoralis minor. The brachial plexus cords are held together with the axillary artery with the axillary sheath. The terminal branches are then given off at the lower outer border of the pectoralis minor.

The ventral rami of the brachial plexus from the fifth and sixth cervical nerve (C5–6) form the *upper trunk*. The *middle trunk* originates from the seventh cervical nerve (C7). The *lower trunk* of the brachial plexus is formed from the eight cervical and the first thoracic spinal nerves (C8–T1). Each of these trunks splits into *anterior* and *posterior divisions*. The division that happens at this point gives an idea of what nerves will innervate the front, and which will innervate the back.

The lateral cord is formed from the *anterior divisions* of the *upper* and *middle trunks*. The *medial cord* is from the *anterior division* of the *lower trunk*. The *posterior cord* arises from the *three posterior divisions*. The

Table 2.2 The Anatomical Location of Each Part of the Brachial Plexus (Roots (Ventral Rami), Trunks, Divisions, Cords, and Terminal Branches)	
Part of Brachial Plexus	Location
Roots and trunks	Neck, close to the subclavian artery
Divisions	Posterior to the clavicle
Cords and terminal branches	Axilla, close to the axillary artery

lateral, medial, and posterior cords tend to lie posterior to the axillary artery, and then wind round the artery to lie either lateral, medial, or posterior relative to this vessel. At the lower outer border of the pectoralis minor, the lateral, medial, and posterior cords split into their terminal branches. Due to the complex formation of the brachial plexus, the terminal branches tend to have several spinal nerves contributing to each of the final branches. Table 2.2 provides a broad outline of where each region is found anatomically. Table 2.3 provides an easy summary to identify what contributes to each peripheral (terminal) nerve from the roots, trunks, divisions, and cords.

Table 2.3 Formation of the Terminal Branches (Peripheral Nerves)

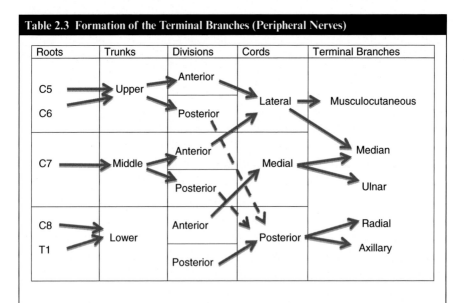

Note: The branches of the ventral rami of C5–T1 forming the brachial plexus include the dorsal scapular nerve, long thoracic nerve, and also small branches to the scalene and longus colli muscles. The table also highlights what forms the cords, divisions and trunks, from the ventral rami of C5–T1.

2.2.1 Dorsal Scapular Nerve

The dorsal scapular nerve originates primarily from the fifth cervical vertebral level nerve (C5). It pierces the scalenus medius. This muscle arises from the posterior tubercles of the transverse processes of the cervical vertebrae and passes to the impression on the superior surface of the first rib. Collectively, along with scalenus anterior and posterior, these muscles flex the cervical part of the vertebral column laterally. The scalene muscles can act on inspiration in normal breathing, but become more pronounced in their activity during stronger voluntary expiration, and can also help in coughing.

The dorsal scapular nerve, after piercing the scalenus medius, then runs deep to the levator scapulae. This muscle originates from the posterior tubercles of the transverse processes from the first to the fourth cervical vertebrae. It then runs to the scapula (medial border) at the spine and slightly above the spine of this bone. The levator scapulae are responsible for elevation of the scapula and can work alongside trapezius in raising the shoulders. It can also work in tandem with the rhomboid muscles helping retract and fixing of the scapula.

After running deep to the levator scapulae, the dorsal scapular nerve then supplies the rhomboid major and minor muscles. The rhomboid major muscle originates from the spines and the supraspinous ligaments of the second to the fifth thoracic vertebrae. It then inserts onto the medial side of the scapula, inferior to the spine. Rhomboid minor arises from the spines of the seventh cervical vertebra, the first thoracic vertebra, and the inferior part of ligamentum nuchae. It then runs to insert onto the medial border of the scapula at the root of the spine. All of the features of these muscles and their respective innervations are shown in Table 2.4.

2.2.1.1 Clinical Examination

When undertaking any clinical history taking or examination, you should always do the following, and follow a logical and systematic format:

1. Introduce yourself to the patient.
2. Advise them of what position you hold, for example, student, specialty grade, consultant, etc.

Table 2.4 The Insertion Point, Origin, Functions, and Innervation of the Muscles Related to the Pathway of the Dorsal Scapular Nerve

Muscle	Origin	Insertion	Actions	Nerve Supply
Scalenus medius	Posterior tubercles of the transverse processes of the cervical vertebrae (C1–C7)	Superior surface of the first rib	Flexion of the cervical part of the vertebral column laterally	Ventral rami of the cervical nerves
Levator scapulae	Posterior tubercles of the transverse processes	Medial border of spine of scapula	Elevates scapula	C3–C4
Rhomboid major	Spinous processes and the supraspinous ligaments of T2–T5	Medial side of the scapula	Retraction of scapula and depression of glenoid cavity	Dorsal scapular nerve (C5)
Rhomboid minor	Spinous processes of C7 and T1	Medial border of scapula	Retraction and rotation of the scapula	Dorsal scapular nerve (C5)

3. Explore your reason for consulting with them, or to find out why they have presented to you.
4. Always take a thorough and detailed history that will be guided by the presenting signs and symptoms.
5. When examining the patient, always tell them what you will ask them to do, or what region of the body you will be examining, with specific instructions, and ensure they give consent.

A detailed examination and history taking should be completed as described in Chapter 1.

The rhomboids and the levator scapulae should be tested together as a functional group.

The rhomboids are involved in adduction (retraction) of the scapula and elevation or downward rotation to allow the glenoid cavity to face caudal. The rhomboids also help fix the scapula to the wall of the thorax.

The levator scapulae, if fixed, not only elevate the scapula but also help with downward rotation to allow the glenoid cavity to face caudal. When its insertion is fixed and acts on one side, it also flexes the cervical vertebrae laterally and rotates it toward the same side. When the insertion of the levator scapulae is fixed and acting on both sides, it could also help with extension of the cervical vertebral column.

On clinical examination, the following should be followed as a suggestion to examine the rhomboids and the levator scapulae:

1. *Palpate*: With the patient standing or sitting, feel the muscle bellies over the back. These will be found medial to the scapula's vertebral border. The fibers of the levator scapulae will be found running diagonally from the transverse processes of the vertebrae to the medial border of the scapula. The rhomboids will be felt inferior to the levator scapulae. Also observe for any asymmetry between the left and right sides; note any atrophy or muscle fasciculations while doing this.
2. *Power of the muscles*: With the patient in the prone position, the arms should be at the edge of the examining couch/bed. With the elbow flexed, the humerus should be adducted toward the side of the body but in slight lateral rotation and also extension. Then, stabilize the shoulder and check the resistance with adduction and downward rotation of scapula by the application of pressure against the arm. This will help assess the rhomboids, ability to hold the scapula in the test position.

2.2.1.2 Clinical Applications
2.2.1.2.1 Dorsal Scapular Nerve Syndrome
Dorsal scapular nerve syndrome typically presents with a weakness of the levator scapulae and the rhomboid muscles, and results in a winged scapula. A winged scapula is where the scapula protrudes from the patient's back and can affect the ability to lift objects or pull and push. The winging of the scapula seen with dorsal scapular nerve syndrome is not as severe as seen as injury or paralysis of the serratus anterior muscle.

Typically, patients with dorsal scapular nerve syndrome will present with a history of abnormal movement of the shoulder joint, and perhaps reduced movement. There may also be pain present over the shoulder region.

On examination, there may be mild winging of the scapula with the lower and medial borders of the scapula raised from the chest wall. The patient will find it difficult to bring the scapulae together. Also, moving the arm forward on elevation will result in the medial border of the scapula being lifted and pulling on the inferior angle of this bone. It is essential to test the full range of movements of the scapula when undertaking clinical examination.

Treatment of dorsal scapular nerve syndrome tends to be conservative but if surgery is contemplated, an MRI scan should be arranged (Pećina et al., 2001).

2.2.1.2.2 Dorsal Scapular Nerve Entrapment

Isolated dorsal scapular nerve injuries are not common and it may become injured if there is hypertrophy of the scalenus medius muscle, or perhaps with shoulder dislocation. This could be found in some sports like judo, or indeed in bodybuilders who may overstretch the scalenus medius during neck flexion and raising shoulders forcefully if the muscle is hypertrophied (Akuthota and Herring, 2009).

Again, this could present with pain in the region of the shoulder or neck, or simply only at the medial border of the scapula. It may also present with winging of the scapula and weakness of the rhomboids. This can also be tested for by having the patient lower their arms from the forward-flexed position. If the rhomboids and perhaps levator scapulae are paralyzed or reduced in function, the examiner will be able to easily place their fingers under the medial border of the scapula.

2.2.2 Long Thoracic Nerve

The long thoracic nerve originates from the fifth to the seventh cervical nerves (i.e., C5–C7) and is said to have almost 1800 nerve fibers (Narakas, 1989). The upper two roots (C5–C6) pierce through the scalenus medius and the lowest (C7) passes anterior to that muscle. Occasionally, the long thoracic nerve also receives contributions from the fourth cervical nerve (Hamada et al., 2008). The long thoracic nerve then passes posterior to the brachial plexus as well as the first part of the axillary artery. The first two roots of the long thoracic nerve then pass toward the scalenus medius, enter in, and join together, accompanied by the seventh cervical nerve root. The long thoracic nerve then passes in an almost vertical direction as it passes down to the serratus anterior muscle and passes to the external surface of the serratus anterior to innervate that muscle with its numerous branches. Ebreheim et al. (1998) showed that the long thoracic nerve, as it passes downward and

Table 2.5 The Origin, Insertion, and Functions of the Serratus Anterior

Origin	Insertion	Function
Upper ribs	Scapula at:	
	Inferior angle	Protraction of the scapula
	Superior angle	Inferior angle moves laterally
	Medial border	Plays an important role in abduction of the arm and elevation of the arm above the horizontal

slightly backward from the point of appearance to the posterior angle of the second rib, is found to take approximately a 30° posterior angle relative to the anterior axillary line. It then has been noted to pass downward between the posterior and middle axillary lines (Ebreheim et al., 1998) Table 2.5 summarises the serratus anterior muscle for its origin, insertion, and function.

The long thoracic nerve is also called the *external respiratory nerve of Bell*. This was named after Sir Charles Bell who first described it in relation to one case where a patient had a fracture of the sixth and seventh cervical vertebrae. This patient had paralysis of the whole body below that level of injury and the "serratus magnus was called into action with every extraordinary effort of breathing, but not by any efforts at voluntary motion" (Willison, 2015). This was when the long thoracic nerve, or external respiratory nerve of Bell, was related to the function of supplying the serratus anterior (magnus).

2.2.2.1 Clinical Examination
When undertaking any clinical history or examination, you should always do the following, and follow a logical and systematic format:

1. Introduce yourself to the patient.
2. Advise them of what position you hold, for example, student, specialty grade, consultant, etc.
3. Explore your reason for consulting with them, or to find out why they have presented to you.
4. Always take a thorough and detailed history that will be guided by the presenting signs and symptoms.
5. When examining the patient, always tell them what you will ask them to do, or what region of the body you will be examining, with specific instructions and ensure they give consent.

A detailed examination and history taking should be completed as described in Chapter 1.

In testing for the long thoracic nerve, the *"serratus wall test"* should be performed. First observe the back of the patient to identify if there is any asymmetry between the left and right sides, identify and note if there is any atrophy of the muscle, or fasciculations.

Then, the patient should be asked to face a wall and approximately at 2 ft. away from it, to push against the wall with the palms flat against the wall at waist level. If there is damage to the long thoracic nerve, and therefore pathology of the muscle it supplies – the serratus anterior – there will be winging of the scapula. Indeed, if the patient is able to, winging of the scapula is also noticeable when performing "push-up" exercises (Narakas, 1989).

2.2.2.2 Clinical Applications

Injury to the long thoracic nerve may occur for a variety of reasons. It could arise as a complication of surgery (mastectomy, cardiac surgical procedures), trauma to the neck and chest, and could also result from some sports like archery, tennis, wrestling and weightlifting (Sunderland, 1978; Narakas, 1989; Wiater and Flatow, 1999; Pećina et al., 2001).

In addition to this, damage to the long thoracic nerve has been documented with surgical resection of ribs, surgical management of patients with pneumothorax, surgical sectioning or division of the scalene muscles, mastectomy with axillary clearance, or anesthesia of the infraclavicular plexus (Kauppila and Vastamaki, 1996).

Indeed, surgery for thoracic outlet syndrome (see Section 2.2.4) caused by an accessory first rib has also been noted (Roos, 1971; Wood et al., 1988; Machleder, 1991). This can happen as the scalenus medius has to be detached from the first rib, and the long thoracic nerve damage (palsy) tends to happen because of a retractor stretching the nerve when exposing the ribs.

Much of the sports-related damage to the long thoracic nerve results from the continual stretching of the nerve. Gregg et al. (1979) demonstrated in the cadaver that the long thoracic nerve could actually be stretched to twice its normal length when the arm was raised and the head rotated to the opposite side. Bora et al. (1976) had shown that a nerve can cope with being stretched to approximately a 10% increase in its resting length, but further stretch would result in damage. Therefore, with the possibility that the long thoracic nerve could be stretched to double its resting length, and the fact that the nerve is enclosed tightly in the axillary sheath, with a fixed distal point of attachment, shows how this nerve can therefore be easily put at risk from neuropraxia.

An interesting area related to the long thoracic nerve is that it has been described by Schultes et al. (1999) as being very similar in structure to the spinal accessory nerve. They had stated that the functional deficit associated with a spinal accessory nerve palsy is far greater than that due to paralysis of the serratus anterior muscle via the long thoracic nerve. Therefore, if the spinal accessory nerve was damaged, for whatever reason, the long thoracic nerve could be used as a nerve graft. This would enable normal to near-normal functioning of the trapezius if the long thoracic nerve was substituted for the damaged spinal accessory nerve.

2.2.3 Suprascapular Nerve

The suprascapular nerve originates from the fifth and sixth cervical nerves (i.e., C5–C6). It passes laterally, deep to the omohyoid and trapezius muscles and courses posterior toward the scapular notch, through which it passes, inferior to the transverse scapular ligament. The suprascapular nerve then passes deep to the supraspinatus, and accompanies the suprascapular atery as it winds round the spine (outer border) of the scapula, thus gaining access to the infraspinous fossa.

The suprascapular nerve innervates the shoulder joint as well as the acromioclavicular joint in the supraspinous fossa. In terms of muscular innervation, the suprascapular nerve also innervates the supraspinatus muscle. The suprascapular nerve also passes through the spinoglenoid fossa, or the great scapular notch (connecting the supraspinous and infraspinous fossa), and it terminates in the infraspinatus that it also innervates.

The *supraspinatus* arises from the *medial two-thirds of the supraspinous fossa* and the *overlying fascia*. Its tendon of insertion blends with the capsule of the shoulder joint and is attached to the highest of three facets on the greater tubercle of the humerus. The tendon forms the floor of the subdeltoid bursa and it often shows signs of wear and tear. The supraspinatus is supplied by the suprascapular nerve and this muscle aids the deltoid in abduction of the arm. Both the deltoid and the supraspinatus contract simultaneously when abduction begins. The supraspinatus, the infraspinatus, teres minor, and the subscapularis keep the head of the humerus in place and prevent it from being pulled up against the acromion by the deltoid. If the deltoid is paralyzed, the supraspinatus cannot usually fully abduct the arm, and if the supraspinatus is paralyzed, normal abduction is difficult or impossible.

The *infraspinatus* is covered in its upper part by the deltoid laterally and the trapezius medially. The muscle arises from the medial two-thirds of the infraspinous fossa and, by a fasciculus that passes downward and laterally (Figure 2.1), from the lower surface of the spine of the scapula. Its tendon, which may show attritional changes, blends with the capsule

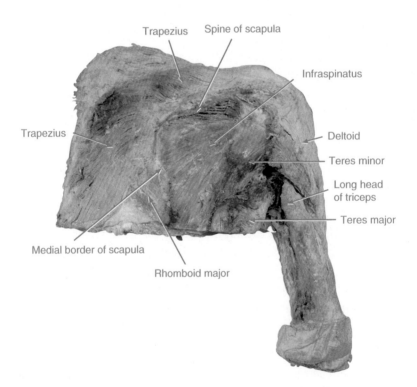

Fig. 2.1. The position of infraspinatus and the teres minor, along with other nearby anatomical structures.

of the shoulder joint and is inserted into the middle facet on the greater tubercle of the humerus. A bursa is usually present between this muscle and the spine of the scapula near the spinoglenoid notch, and another may be present between the tendon and the joint capsule.

2.2.3.1 Clinical Examination

When undertaking any clinical history or examination, you should always do the following, and follow a logical and systematic format:

1. Introduce yourself to the patient.
2. Advise them of what position you hold, for example, student, specialty grade, consultant, etc.
3. Explore your reason for consulting with them, or to find out why they have presented to you.
4. Always take a thorough and detailed history that will be guided by the presenting signs and symptoms.
5. When examining the patient, always tell them what you will ask them to do, or what region of the body you will be examining, with specific instructions and ensure they give consent.

A detailed examination and history taking should be completed as described in Chapter 1.

Specific to the nerve supply of the suprascapular nerve and the muscles it innervates, that is, supraspinatus and infraspinatus, a comprehensive examination of the shoulder joint should be performed.

The shoulder joint and its related movements are a complicated action of four joints. The glenohumeral joint and the *thoracoscapular mechanism* both account for approximately half the total range of movement of the shoulder joint. The sternoclavicular and acromioclavicular joints solely account for a relatively small amount of the shoulder joint movement.

With the scapula in a fixed position, the glenohumeral joint accounts for the following:

- 90° flexion
- 90° abduction
- 30° extension
- 90° medial (internal) and lateral (external) rotation

The other range of movements are when the scapula becomes recruited and this is referred to as the *thoracoscapular mechanism*. When the elbow is in the flexed position to a right angle the forearm can be used as an indicator of rotation. When the arm is in the neutral position, abduction can only proceed to 90°. When the arm is externally (laterally) rotated, further abduction can then be obtained. When the arm is by the side in a neutral position, and the elbow flexed to 90° the forearm points medially by approximately 20°. Then, rotation to 90° in either direction can result.

The *rotator cuff* is the tendons of the supraspinatus and infraspinatus as well as the teres minor (see Figure 2.1). Subscapularis also plays a role in the *rotator cuff muscles*. The cuff forms a hood over the head of the humerus, keeps it within the glenohumeral joint, and allows for the initiation of abduction. The rotator cuff is kept separate from the acromion (inferior apart) by the subacromial bursa. The rotator cuff, the supraspinatus tendon, and associated subacromial bursa, is the most common part of the shoulder to suffer from pathology.

1. *Inspection*: Observe if there is any muscular atrophy. If there is a chronic restriction of movement due to, for example, a frozen shoulder (adhesive capsulitis) or arthritic changes, there will be marked muscular atrophy. Synovial effusion is difficult to identify in the shoulder joint due to the deltoid overlying the joint. Always compare both sides of the shoulder joint and back.
2. *Palpation*: With rotator cuff disease, there may be tenderness over the subacromial space and also the tip of the shoulder (supraspinatus tendinitis), over the bicipital groove anteriorly (bicipital tendinitis) and, if acute effusion and synovitis is present, there may be swelling and pain over the anterior capsule. Note any points or location of tenderness especially over the anterior region, sternoclavicular and acromioclavicular joints.
3. *Movement*: This is best done by standing behind the patient and observing the whole range of movements of the shoulder joint. The examiner should ask the patient to put their hands at the bottom of their neck, with their elbows pointing out laterally. Then, the patient can be asked to put their upper limbs down by their side, and then to reach behind the back to an area between the two scapulae if possible.

Then with the patient's upper limbs by their sides and the elbow flexed to 90°, observe the range of movement in rotation. A loss of external rotation would point toward a glenohumeral joint issue.

In examining the glenohumeral joint, the scapula must be stopped from moving. This can be done by the examiner placing a hand over the tip of the scapula, or by putting a restraining hand over the top of the acromion. With the arm by the patient's side, ask them to then flex their shoulder anteriorly, and should normally reach 90°. Then, abduction should be tested out to 90°. If the patient finds abduction difficult, the examiner should then test for the rotator cuff muscles as described further. If the patient is not able to perform abduction, the examiner can perform this passively, and if this becomes uncomfortable or painful for the patient, then testing for painful arc syndrome should be performed.

2.2.3.1.1 Rotator Cuff Muscle Testing

1. With the arm by the side, ask the patient to abduct their arm against resistance. If the supraspinatus tendon is inflamed, this will be painful for the patient to carry out. If there is rupture of the rotator cuff, initiation of abduction will be impossible.
2. If a rupture is suspected, start abduction for the patient (i.e., passively) out to about 30–45°. The patient should then be able to complete the rest of abduction from this aided position of the shoulder.
3. While restraining the shoulder girdle, test for impingement pain by bringing the patient's arm into flexion greater than 90°.
4. Rotation of the shoulder girdle internally (medially) in the 90° flexed position to exacerbate this impingement pain. This is referred to as *Hawkins sign*.
5. Finally, ask the patient to bring the arm into full elevation and then get the patient to bring their arm downward to elicit a painful arc.
6. If there is a complete rupture of the supraspinatus tendon, the patient will be unable to initiate abduction. If the deltoid muscle is intact with rupture of the supraspinatus tendon, the patient will only appear to shrug when trying to abduct the shoulder.

2.2.3.1.2 Painful Arc Syndrome Test

1. Pain will be noted at approximately 45–140° of abduction and raising the arm further into elevation.
2. If the patient is unable to do this due to acute pain, impingement pain can be demonstrated by rotation of the flexed arm. In the dependent position, this will be pain free, thus confirming no glenohumeral pathology.
3. If there is a painful arc or impingement pain, it suggests subacromial bursitis, incomplete tear of the supraspinatus tendon, or supraspinatus tendinitis.

2.2.3.2 Clinical Applications

2.2.3.2.1 Rotator Cuff Injury

Injury to the rotator cuff muscles can occur in those individuals who repeatedly undertake overhead movements in their occupation, or sport participation. This typically includes painters and decorators, tennis and baseball players, and carpenters. Two main problems occur with the rotator cuff muscles – tendinitis or tears.

1. *Tendinitis*: This is inflammation of the tendons forming the rotator cuff muscles, and can result if the arms are kept in the same position for long periods of time, for example, office-based staff that use computers frequently, or hairdressers. It could also happen if the patient sleeps on the same arm each night, or participates in sports like tennis, baseball, swimming, or weightlifting.
2. *Tears of the rotator cuff muscles*: Tears to the rotator cuff muscles may happen either acutely, or develop in longer term. For the acute situation, this may be due to direct trauma to the shoulder joint and with chronic conditions, it can also present with impingement pain, as previously described. The tear may be partial or complete, and can still present with microscopic tears to the tendons.

Treatment of rotator cuff muscle injury depends on the cause, and it may involve removing the exacerbating factor, like participation in sports, or with conservative measures like ice packs and pain relief medication. For those with more severe signs and symptoms, surgical correction may be needed with physiotherapy input to help strengthen the rotator cuff muscles over a period of time.

2.2.4 Nerve to Subclavius

The nerve to subclavius arises primarily from the fifth cervical vertebral level (C5). It is a very small branch and can arise from the junction between the fifth and sixth cervical vertebral nerves (C5/6). After originating from the cervical nerve, it descends posterior to the clavicle, in front of the brachial plexus, and the third part of the subclavian artery. It terminates in the structure it supplies – subclavius – after passing above the subclavian vein to reach this muscle. The nerve to subclavius also innervates the sternoclavicular joint. The subclavius muscle arises by a tendon from the junction of the first rib with its costal cartilage, and is inserted by muscle fibers into a groove on the lower surface of the clavicle. The subclavius acts to assist in depressing the lateral part of the clavicle. Movement of this region can be assessed during the examination of the shoulder joint. Frequently, the nerve to subclavius also supplies fibers to the phrenic nerve through a communicating branch called the accessory phrenic nerve.

The phrenic nerve is the nerve that innervates the diaphragm via its nerve roots from the fourth and fifth cervical nerves, as well as input from the third cervical nerve (i.e., C3–5). The accessory phrenic nerve arises from that part of the nerve to subclavius. In some instances this contribution can run a separate course into the thorax before joining the phrenic nerve and therefore is termed the accessory phrenic nerve. Occasionally, the phrenic nerve arises in this way. There can be accessory contributions that are not connected at all to the nerve to subclavius.

The accessory phrenic nerve typically passes anterior to the subclavian vein, whereas the phrenic nerve will run posterior to this vein. If an accessory nerve is present, sectioning of the phrenic nerve, or injury to it in the neck, will fail to paralyze the corresponding half of the diaphragm completely. The accessory nerve will therefore provide some innervation to the diaphragm to aid in its function.

2.2.4.1 Clinical Examination

When undertaking any clinical history or examination, you should always do the following, and follow a logical and systematic format:

1. Introduce yourself to the patient.
2. Advise them of what position you hold, for example, student, specialty grade, consultant, etc.

3. Explore your reason for consulting with them, or to find out why they have presented to you.
4. Always take a thorough and detailed history that will be guided by the presenting signs and symptoms.
5. When examining the patient, always tell them what you will ask them to do, or what region of the body you will be examining, with specific instructions and ensure they give consent.

A detailed examination and history taking should be completed as described in Chapter 1. The following demonstrates a broad outline of how to examine the shoulder joint, and the clavicle, and the sternoclavicular joint can be observed during these movements.

1. *Inspection*: Observe if there is any muscular atrophy. If there is a chronic restriction of movement due to, for example, a frozen shoulder (adhesive capsulitis) or arthritic changes, there will be marked muscular atrophy. Synovial effusion is difficult to identify in the shoulder joint due to the deltoid overlying the joint. Always compare both sides of the shoulder joint and back.
2. *Palpation*: With rotator cuff disease, there may be tenderness over the subacromial space and also the tip of the shoulder (supraspinatus tendinitis), over the bicipital groove anteriorly (bicipital tendinitis) and, if acute effusion and synovitis is present, there may be swelling and pain over the anterior capsule. Note any points or locations of tenderness, especially over the anterior region, sternoclavicular, and acromioclavicular joints.
3. *Movement*: This is best done by standing behind the patient and observing the whole range of movements of the shoulder joint. The examiner should ask the patient to put their hands at the bottom of their neck, with their elbows pointing out laterally. Then, the patient can be asked to put their upper limbs down by their side, and then to reach behind the back to an area between the two scapulae if possible.
   Then, with the patient's upper limbs by their sides and the elbow flexed to 90° observe the range of movement in rotation. A loss of external rotation would point toward a glenohumeral joint issue.

2.2.4.2 Clinical Applications

Rarely, there can be anomalous or accessory muscles within the root of the neck. One such muscle is the *subclavius posticus*. This muscle has been found to attach to the superior border of the scapula, situated lateral to the attachment of the lower belly of the omohyoid. The proximal part of this muscle was found to originate from the costal cartilage of the first rib, and is typically found deep to the subclavius muscle. In one such study, it was described measuring 11.5 cm in total length and having a flat, triangular belly that was 6 cm long and 1.1 cm wide; this muscle as receiving its innervation from the suprascapular nerve, although others have described it as receiving its innervation from the nerve to subclavius.

It has been suggested that this supernumerary muscle (subclavius posticus) could well present as a possible cause of the thoracic outlet syndrome.

2.2.4.2.1 Thoracic Outlet Syndrome

The brachial plexus and the subclavian artery and vein pass through an extremely narrow space between the clavicle and the first rib. This space can become narrowed for a variety of reasons and can include the following:

1. *Congenital*: This may result in an extra rib arising from the seventh cervical vertebra. Typically the ribs will originate from the first to twelfth thoracic vertebrae. It can occur on one or both sides. In general, they are asymptomatic, but can present as thoracic outlet syndrome. Other causes of congenital reasons for thoracic outlet syndrome can be from muscle variations either of the scalenus anterior, or, indeed, a subclavius posticus. There may also be additional fibrous material within the neck resulting in the obstruction.
2. *Trauma*: This may result from trauma to the neck or also from repetitive strain. Indeed, for those habitually involved with their hands above their heads when working (e.g., window cleaners, laborers), it may exacerbate the symptoms.
3. *Tumors within the neck*

Patient's with thoracic outlet syndrome will typically present with pain in the upper limb, of variable distribution. It can affect the digits, hand or arm, and forearm. Pain can also present in the axilla, superior aspect of the back or over the pectoral region inferior to the clavicle. In addition to the pain, the patient will typically present with a discoloration of the upper limb, generally the hand, and it can be colder to the touch compared to the unaffected side.

If a patient has a swollen, painful and blue tinged upper limb, especially after strenuous exercise, it may indicate compression of the subclavian vein, thoracic outlet compression, and thrombosis. This is referred to as *Paget–Schroetter syndrome*, but can also be called effort-induced thrombosis.

The treatment of thoracic outlet syndrome can range from conservative measures, to removal of a tumor or extra (cervical rib) if the symptoms are extreme. Therefore, the treatment depends on the cause of the thoracic outlet syndrome.

2.2.4.2.2 Erb's Point

The nerve to subclavius is from the fifth and sixth cervical nerves, and Erb's point is where the upper part of the brachial plexus is found approximately 2–3 cm superior to the clavicle. From these nerves the deltoid, brachialis, brachioradialis, brachialis, and the biceps are also supplied from C5 to C6.

Injuries to the upper part of the brachial plexus (i.e., C5–C6) result in paralysis of the shoulder and the arm. These types of injuries classically happen when the angle between the head and neck are increased, for example, childbirth, fall from a motorbike or horse, any mechanism resulting in a fall to the top of the shoulder, or indeed from repetitive damage due to carrying heavy backpacks. Paralysis of the deltoid, brachialis, brachioradialis, and the biceps would result in an *Erb–Duchenne palsy*. Typically this presents with:

1. Adducted shoulder,
2. Medial rotation of the arm, and
3. Extension of the elbow.

This appearance is referred to as the "waiter's tip" position due to the positioning of the upper limb with the aforementioned features.

2.2.5 Lateral Pectoral Nerve

The pectoral nerves supply the pectoralis major and minor. The lateral pectoral nerve arises from the fifth through to the seventh cervical nerves (C5–C7) and is the larger of the two pectoral nerves. It arises from the lateral cord or from the anterior divisions of the upper and middle trunks. The lateral pectoral nerves then pass anterior to the axillary vein and artery anteriorly and it pierces the clavipectoral fascia, terminating in the muscle it supplies – the *pectoralis major*. They send a loop across the first part of the axillary artery to join the medial pectoral nerves and, through this, contribute fibers to the pectoralis minor muscle. The origin, insertion, innervation, and functions of the pectoralis major muscle are given in Table 2.6.

2.2.6 Medial Pectoral Nerve

The medial pectoral nerves arise from the eighth cervical nerve and the first thoracic nerves (C8–T1). The medial pectoral nerve comes from the medial cord or from the lower trunk. These pass anterior between the axillary artery and vein, and then pierce the *pectoralis minor* to supply it. Some branches can pass around the lower border of the pectoralis minor to then innervate the pectoralis major. The origin, insertion, innervation, and functions of the pectoralis minor muscle are given in Table 2.6.

Table 2.6 The Origin, Insertion, Innervation, and Functions of the Pectoralis Major and Minor Muscles

	Proximal Attachment	Distal Attachment	Nerve Supply	Function
Pectoralis major	Anterior surface of the medial half of the clavicle. Anterior surface of the sternum. First six costal cartilages. Aponeurosis of the external oblique muscle	Crest of the greater tubercle of the humerus	Lateral pectoral nerve (C5–C7) and to a lesser extent the medial pectoral nerve (C8–T1)	Adduction of the arm. Medial rotation. Clavicular part – flexion of the arm. Sternocostal part – depression and adduction of the arm and shoulder
Pectoralis minor	Anterior external aspect from 2nd to 5th ribs	Coracoid process	Medial pectoral nerve (C8–T1) and to a lesser extent the later pectoral nerve (C5–C7)	Depression of the shoulder

2.2.6.1 Clinical Examination

When undertaking any clinical history or examination, you should always do the following, and follow a logical and systematic format:

1. Introduce yourself to the patient.
2. Advise them of what position you hold, for example, student, specialty grade, consultant, etc.
3. Explore your reason for consulting with them, or to find out why they have presented to you.
4. Always take a thorough and detailed history that will be guided by the presenting signs and symptoms.
5. When examining the patient, always tell them what you will ask them to do, or what region of the body you will be examining, with specific instructions and ensure they give consent.

A detailed examination and history taking should be completed as described in Chapter 1. When undertaking examination of the pectoralis major and minor, this will be covered in an appropriate shoulder examination, but bearing in mind the functions of these two muscles, then these actions should be tested, that is, adduction of the arm, medial rotation, and flexion of the arm.

2.2.6.2 Clinical Applications

2.2.6.2.1 Injury of Pectoralis Major

Injury to the pectoralis major tends to occur at the musculotendinous junctions perhaps from excessive bench-press exercises. Tears are not common of the pectoralis major muscle, but may warrant surgical intervention, especially in athletes.

2.2.6.2.2 Poland Syndrome

Poland syndrome is a rare condition that presents with an absent sternocostal part of the pectoralis major muscle. Typically it affects 1–3 per 100,000 newborn children (Genetics Home Reference, 2015). Other muscles of the chest may be absent or underdeveloped, with less subcutaneous adipose tissue over the pectoralis major, and perhaps nipple abnormalities.

In addition, hand abnormalities are typically seen with Poland Syndrome. This can include vestigial fingers, fusion of the fingers (*syndactyly*), or shorter-than-usual fingers (*brachydactyly*).

The severity of the condition will dictate the symptoms of the patient. There will be limitation in medial rotation of the humerus and adduction due to the lack of the sternocostal head of the pectoralis major muscle.

2.2.7 Musculocutaneous Nerve

The musculocutaneous nerve originates from the fifth to seventh cervical nerves (C5–C7), typically from the lateral cord and then pierces the coracobrachialis (see Figure 2.2). Sometimes it can also convey some or all of the lateral head of the median nerve and give these fibers to the medial head of the medial nerve with a communication in the arm.

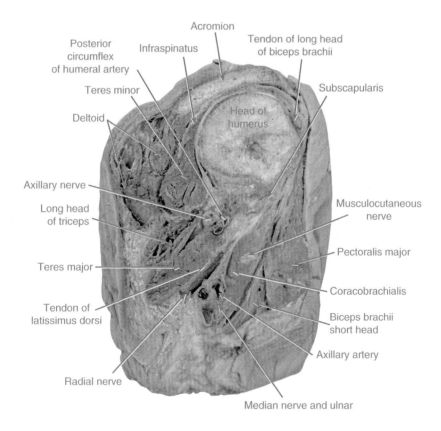

Posterior   Anterior

Fig. 2.2. The musculocutaneous nerve passing through the coracobrachialis.

Table 2.7 The Origin, Insertion, and Functions of Each of the Muscles Supplied by the Musculocutaneous Nerve

Muscle	Origin	Insertion	Functions
Coracobrachialis	Coracoid process	Middle third of medial side of humerus	Flexion of arm Adduction of humerus
Biceps brachii	Short head – coracoid process of scapula Long head – supraglenoid tubercle	Radial tuberosity Bicipital aponeurosis	Flexion of the elbow Flexion of the shoulder Abduction of the shoulder Supination at the radioulnar joint
Brachialis	Distal two-thirds of anteromedial and anterolateral surfaces of the humerus	Capsule of the elbow joint and anterior surface of the coronoid process and tuberosity of the ulna	Flexion of elbow

Therefore, the lateral cord can divide lower than usual. In some cases, the musculocutaneous nerve can travel with the lateral head of median nerve and subsequently be given back as a communication to the musculocutaneous nerve. Ultimately, the musculocutaneous nerve innervates the flexor muscles (described further) on the anterior aspect of the arm, skin on the lateral side of the forearm, and also to the elbow joint. The innervation of coracobrachialis can come from the lateral cord separately, rather than from the musculocutaneous nerve itself.

If the musculocutaneous nerve arises from within the axilla, it usually pierces the coracobrachialis. It will then pass inferiorly between the biceps superficially and the brachialis deeper, thus reaching the lateral aspect of the arm.

The musculocutaneous nerve innervates the coracobrachialis, biceps, and the brachialis as well as the elbow joint (see Table 2.7). It will then become the *lateral antebrachial cutaneous nerve*, or the *lateral cutaneous nerve of forearm* (Figure 2.3) that then goes through the fascia lateral to the tendon of the biceps tendon, just superior to the elbow. This will then divide into an *anterior* (*volar*) and *posterior* (*dorsal*) branch that may lie posterior to the cephalic vein. The anterior branch will innervate the skin on the anterior aspect of the radial side of the forearm as far as the thenar eminence. The posterior branch, however, innervates the skin on the lateral and posterolateral region of the forearm as far as the wrist. These branches will innervate a variable amount of skin on the dorsal aspect of the hand.

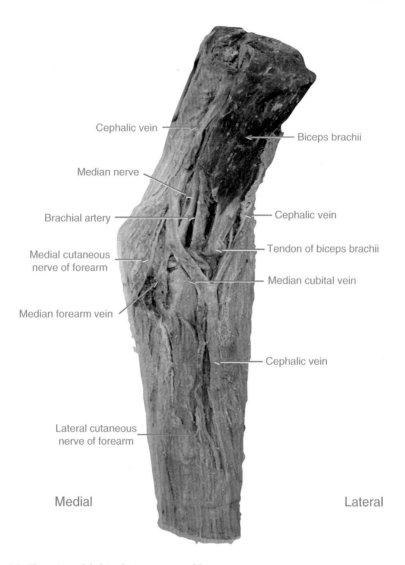

Cephalic vein

Biceps brachii

Median nerve

Brachial artery

Cephalic vein

Tendon of biceps brachii

Medial cutaneous
nerve of forearm

Median cubital vein

Median forearm vein

Cephalic vein

Lateral cutaneous
nerve of forearm

Medial

Lateral

Fig. 2.3. The position of the lateral cutaneous nerve of forearm.

2.2.7.1 Clinical Examination

When undertaking any clinical history or examination, you should always do the following, and follow a logical and systematic format:

1. Introduce yourself to the patient.
2. Advise them of what position you hold, for example, student, specialty grade, consultant, etc.

3. Explore your reason for consulting with them, or to find out why they have presented to you.
4. Always take a thorough and detailed history that will be guided by the presenting signs and symptoms.
5. When examining the patient, always tell them what you will ask them to do, or what region of the body you will be examining, with specific instructions and ensure they give consent.

A detailed examination and history taking should be completed as described in Chapter 1. Examination for the functioning of the biceps brachii, brachioradialis, and the coracobrachialis should be done by determining the power and functioning of the range of movements as described in Table 2.7.

Assessment of the shoulder joints should be performed as detailed previously in Sections 2.2.3 and 2.2.4. In addition to this, the elbow joint should be assessed as follows:

1. *Inspection*: Inspection of the elbow joints should be done by standing behind the patient and observing the elbows in the fully extended position. Note any abnormality or asymmetry.
2. *Palpation*: When feeling the elbow joint, be careful not to hurt the patient, especially if there is localized tenderness. Identify any nodules, points of tenderness or any obvious signs of inflammation.
3. *Movement*
   a. The range of flexion–extension and pronation–supination should be assessed with active movements.
   b. With the elbows flexed at 90° and elbows by the patient's side, pronation and supination should be assessed. If this is not done, abduction and rotation of the shoulder can mislead the examiner for the appearance of pronation and supination.
   c. The examiner should gently palpate the elbow joint during flexion and extension, and also over the head of the radius during pronation and supination to identify if crepitus is present.
   For completeness in the examination of the elbow joint, you must test for epicondylitis.
   a. You must ask the patient to tightly grip their hand when the elbow if fully extended and also when partly flexed. If the patient has tennis elbow, the former maneuver will be painful, and the second

not so much, or not at all. If the tennis elbow is severe, the patient may actually find it difficult simply extending the elbow.

b. Then, place the patient's arm into the "waiter's tip" position, that is, full-elbow extension, pronation of the forearm, and wrist flexed. Then, ask the patient to attempt to extend their fingers against resistance. This will result in exacerbation of the pain at the insertion point of the extensors.

c. In assessing medial epicondylitis, ask the patient to put their arm in the extended position and supinate their forearm. Then, ask the patient to attempt to flex their wrist or fingers against resistance. This will result in pain at the flexor insertion, and can be useful in assessing medial epicondylitis.

2.2.7.2 Clinical Applications

Isolated injury to the musculocutaneous nerve is rare. It may, however, be damaged at the shoulder joint, resulting in a large brachial plexus injury. It may also be compressed by the bicipital aponeurosis and tendon against the brachialis fascia. This would result in anesthesia below the elbow joint on the lateral side of the forearm.

Typically, an injury to the musculocutaneous nerve would result in poor supination, weakened flexion of the elbow joint, and also parasthesia over the lateral aspect of the forearm. In addition, and on examination, the biceps tendon jerk reflex may be absent. Please refer to Chapter 1 for further details on how to carry out this examination. There may be muscle wasting or fasciculations as this would represent a lower motor-neuron lesion, discussed in Chapter 1.

The treatment of injury to the musculocutaneous nerve really depends on the cause, and it may be necessary for surgical intervention.

2.2.8 Median Nerve

The median nerve originates from primarily the sixth cervical nerve to the first thoracic nerve, that is, C6–T1, though sometimes can have input also from the fifth cervical nerve. Therefore, the median nerve can originate from C5 to T1. The median nerve is formed on the lateral aspect of the axillary artery by heads derived from the medial and lateral cords from the brachial plexus. It then continues on the lateral side of the brachial artery. At approximately the middle of the arm, the median

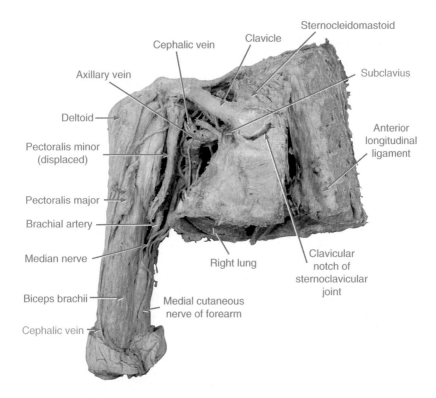

Fig. 2.4. The location of the median nerve within the arm.

nerve passes anterior to the brachial artery (Figure 2.4), but sometimes courses posterior to this vessel.

The median nerve then courses medial to the brachial artery. In the cubital fossa, the median nerve lies behind the median cubital vein and under the bicipital aponeurosis, providing a branch to the elbow joint. The nerve then leaves the cubital fossa typically by passing between the two heads of pronator teres. It is separated from the ulnar artery by the deep or ulnar head of that muscle.

The nerve then courses down the middle of the forearm to the mid-point between the styloid processes. Just prior to passing deep to the flexor retinaculum, it lies superficial between the tendons of the flexor carpi radialis and the palmaris longus. The median nerve then passes deep to the tendinous arch connecting the two heads of flexor digito-rum superficialis. The nerve generally stays under the flexor digitorum

Table 2.8 The Innervation of Structures by the Median Nerve, Both Motor and Sensory		
Median nerve branches			
Forearm		
Muscular branches	Pronator teres Flexor carpi radialis Palmaris longus Flexor digitorum superficialis	
Anterior interosseous nerve	Flexor digitorum profundus (lateral part) Flexor pollicis longus	
Palmar cutaneous branch	Lateral aspect of the palm (not digits)	
Hand		
Lateral division	Muscular branch	Abductor pollicis brevis Flexor pollicis brevis Opponens pollicis
	Palmar digital nerves	All of the thumb Lateral side of the index finger
Medial division	Medial side of index finger and middle finger, and lateral aspect of the ring finger First and second lumbricals	

superficialis and on the flexor digitorum profundus until it reaches the wrist joint.

At the wrist, the median nerve becomes superficial by passing between the flexor carpi radialis and the flexor digitorum superficialis. When present, the palmaris longus partly covers it. The median nerve then enters the hand by passing through the carpal tunnel behind the flexor retinaculum, but in front of the flexor tendons. The median and ulnar nerves may communicate with each other in the forearm, and this typically occurs over the flexor digitorum profundus.

2.2.8.1 Branches of the Median Nerve
The median nerve gives off a variety of branches in its journey through the arm and forearm to the hand. These are summarized in Table 2.8.

2.2.8.2 Carpal Tunnel
The carpal tunnel is an osteofibrous canal situated at the wrist joint. The carpal tunnel is bounded by the carpal bones deeply, thus forming its floor, and the flexor retinaculum superficially, forming its roof. The flexor retinaculum attaches onto the pisiform, tuberosity of the scaphoid, and the trapezium and the hook of the hamate bone. It is divided into

two layers at the radial side due to the tendon of the flexor carpi radialis. Therefore, it contains a deep and superficial layer. The carpal tunnel contains the median nerve as well as nine tendons. Therefore, the carpal tunnel contains the following:

1. Median nerve and tendons of:
   a. Flexor digitorum profundus (x4)
   b. Flexor digitorum superficialis (x4)
   c. Flexor pollicis longus

In close contact with the carpal tunnel, though not passing through it, are the tendons of the following muscles:

1. Flexor carpi ulnaris
2. Flexor carpi radialis
3. Palmaris longus

Table 2.9 summarizes the origin, insertion point, functions, and innervations of those muscles described.

2.2.8.3 Anatomical Variants Within the Carpal Tunnel

1. *Median artery*: The hand receives its blood supply from two main arteries – the ulnar and radial arteries. In some people, the median artery, typically from the ulnar artery, may persist after birth, and not regress, as it does usually in the second month of intrauterine life. If a median artery is found, it may present as a contributing factor in the pain related to carpal tunnel syndrome. Carpal tunnel syndrome will be discussed further later in this section on the median nerve.
   If there is a persistent median artery, this can also occur alongside one of the most common variations of the median nerve – a bifid median nerve. If this is the case, the median artery tends to be located between the two branches of the median nerve.
2. *Proximal bifurcation of the median nerve*: Typically the median nerve will bifurcate at the distal border of the transverse carpal ligament. In some instances, it is quoted that the median nerve may bifurcate higher in the forearm and within the carpal tunnel itself. This can be found in approximately 1–3% of individuals who have surgery for carpal tunnel syndrome. The two branches of the median nerve typically are approximately the same size.

Table 2.9 The Muscles Innervated by the Median Nerve, and Related Muscles and Nerves

Muscle	Origin	Insertion	Innervation	Function
Flexor digitorum profundus	Anterior surface of ulna (upper two-thirds to three-fourths) Coronoid process Interosseous membrane	Bases of distal phalanges of the fingers	Medial: Ulnar nerve Lateral: Anterior interosseous nerve (from median nerve)	Flexes distal phalanges, typically with flexion of the middle phalanges by the flexor digitorum superficialis
Flexor digitorum superficialis	Common tendon from the medial epicondyle of humerus Anterior superior border of radius	Middle phalanges of the four fingers, anteriorly	Median nerve	Flexion of the fingers, mainly at the proximal interphalangeal joints
Flexor pollicis longus	Anterior surface of radius (upper two-thirds to three-fourths) Interosseous membrane	Base of distal phalanx (palmar surface)	Anterior interosseous nerve (from median nerve)	Flexion of distal phalanx of the thumb
Flexor carpi ulnaris	Common tendon from the medial epicondyle of humerus Medial surface of the olecranon	Pisiform Hook of hamate and the base of the fifth metacarpal (via the pisohamate and the pisometacarpal ligaments)	Ulnar nerve	Flexion of the hand Adduction of the hand, with extensor carpi ulnaris Steadies pisiform during abduction of the little finger by abductor digiti minimi Synergistic action with flexor carpi radialis steadying the wrist joint during extension of the fingers Steadies hand during extension and abduction of the thumb, working alongside the extensor carpi ulnaris
Flexor carpi radialis	Common tendon from the medial epicondyle of humerus	Anterior aspect of the base of the second and third metacarpal bones	Median nerve	Flexion of the hand Along with the radial extensors, it aids abduction of the hand Synergistic muscle along with flexor carpi ulnaris, it steadies the wrist during extension of the fingers
Palmaris longus	Common tendon from the medial epicondyle of humerus	Flexor retinaculum Apex of palmar aponeurosis	Median nerve	Tenses the palmar aponeurosis on movement of the hand Weak flexor of the hand

3. *Variation in the motor branch of the median nerve*: The motor nerve
   of the median nerve can arise distal or below the transverse carpal
   ligament, or actually through the substance of the ligament. Rarely,
   it can arise on the ulnar side crossing the median nerve to the
   thenar eminence, or can even be superficial to the transverse carpal
   ligament.

4. *Ulnar nerve variation*: Occasionally, rather than traveling through
   Guyon's canal, the ulnar nerve can travel through the carpal
   tunnel, or may communicate higher up with the median nerve in
   the forearm, that is, a *Martin-Gruber anastomosis*. This results
   in a variety of innervation patterns of the muscles of the hand,
   for example. It is due to fibers arising from the median nerve
   then passing to the ulnar nerve. Indeed, if nerve fibers run from
   the ulnar nerve to the median nerve, this is called a *Marinacci
   communication*, otherwise thought of as a *reverse Martin-Gruber
   anastomosis.*

5. *Muscular variants*: The muscles flexor pollicis longus, flexor
   digitorum profundus and the flexor digitorum superficialis all
   arise from the same *mesodermal* tissue. As such, there may exist
   connections between them, and one of the most frequent found is
   a tendinous connection between the flexor pollicis longus and the
   flexor digitorum profundus. This can result in flexion of the distal
   phalanx of the index finger when there is flexion of the thumb. This
   is referred to as *Linburg–Comstock syndrome*.

   Another muscle that exhibits a high degree of variability is the
   palmaris longus. These must be identified and established especially
   early in surgery. One such variant of the palmaris longus is that the
   muscle may pass through the carpal tunnel. The other variation
   that can be seen with this muscle is in relation to its muscle belly.
   Typically, the belly of this muscle is found situated proximally and
   the tendon located distally. Some people have the reverse of this to
   be true. In other words, the muscle belly can be found distally, and
   the tendon is more pronounced proximally. Indeed, if both of these
   variants occur together, that is, reverse location of the muscle belly
   and traveling through the carpal tunnel, it can be a perfect recipe
   for increasing friction at the carpal tunnel between the osteofibrous
   nature of the tunnel, and the pronounced muscle belly going
   through it.

Regarding the flexor digitorum superficialis, there may also exist variations of this muscle. The muscle belly of flexor digitorum superficialis may pass into the carpal tunnel, accessory muscles may be present toward the wrist (e.g. accessory flexor digitorum superficialis passing to the index finger), or an accessory digastric muscle (muscle belly found both at the palm and in the forearm) may be present. Rarely a muscle exists that can originate from the distal radius, on its palmar aspect, cross the pronator quadratus, then pass deep to the flexor retinaculum, and pass to join on to the third and the fourth metacarpal bones. Again, as with so many of these anatomical variations, it could cause carpal tunnel syndrome.

2.2.8.4 Clinical Examination
When undertaking any clinical history or examination, you should always do the following, and follow a logical and systematic format:

1. Introduce yourself to the patient.
2. Advise them of what position you hold, for example, student, specialty grade, consultant, etc.
3. Explore your reason for consulting with them, or to find out why they have presented to you.
4. Always take a thorough and detailed history that will be guided by the presenting signs and symptoms.
5. When examining the patient, always tell them what you will ask them to do, or what region of the body you will be examining, with specific instructions and ensure they give consent.

A detailed examination and history taking should be completed as described in Chapter 1.

Table 2.10 summarizes the nerve roots, muscle supplied, and relevant clinical test to ask the patient to perform in assessing the median nerve.

To test the integrity of the muscles innervated by the median nerve, it is also essential to assess the sensory distribution of the median nerve, that is, all of the thumb, lateral side of the index finger, medial side of the index finger and middle finger, and the lateral aspect of the ring finger. This should be documented according to the International Standards for Neurological Classification of Spinal Cord Injury (ISNCSCI, 2015).

Table 2.10 The Ways to Clinically Examine the Media Nerve, and What to Ask the Patient to Do, and What Muscle is Being Assessed

Nerve Root	Muscle	Ask the Patient to...
C6–C7	Pronator teres	Keep the arm pronated against resistance from the examiner
C6–C8	Flexor carpi radialis	Flexion of the wrist to the radial side
C7–T1	Flexor digitorum superficialis	Resist extension to the proximal interphalangeal joints. The proximal phalanges should be fixed
C8–T1	Flexor digitorum profundus (1st and 2nd)	Resisting extension of the distal interphalangeal joints
C8–T1	Flexor pollicis longus	Resisting extension of the thumb at the interphalangeal joints. The proximal phalanges should be fixed
C8–T1	Abductor pollicis brevis	Thumb abduction
C8–T1	Opponens pollicis	Opposition of the thumb, that is, the patient has to bring the thumb to touch the 5th finger tip
C8–T1	1st and 2nd lumbricals	Extension of the proximal interphalangeal joint against resistance by the examiner with the metacarpophalangeal joint positioned hyperextended

2.2.8.5 Clinical Applications

Perhaps the most common, and widely appreciated condition to affect the upper limb nerves is carpal tunnel syndrome. This is where the median nerve is compressed in the carpal tunnel, as previously described. It results in the patient having numbness and paresthesia in the fingers, typically affecting the thumb and index and middle fingers, though some patients may feel as though their whole hand is affected. Early in the course of this pathology, the symptoms tend to be worse at night and tend to disappear by the morning. If there is progressive worsening of carpal tunnel syndrome, it can result in muscle wasting of the thenar muscles (i.e., abductor pollicis brevis, flexor pollicis brevis, and the opponens pollicis).

There are two simple tests that can be performed in assessing the possibility of carpal tunnel syndrome – Tinel's test and Phalen's sign.

1. *Tinel's test*: This involves lightly tapping over the carpal tunnel at the wrist joint. This tapping will reproduce the shooting sensations and exacerbate the symptoms of the numbness and paresthesia. Try not to do this too often in the examination as it can be rather painful and uncomfortable for the patient.

2. *Phalen's sign*: Phalen's sign is where you should ask the patient to flex their wrist, or passively flex it for them, for a minute or so, or until it is uncomfortable for the patient. If it is positive, it will result in the appearance of the symptoms, or worsening of them. Do not perform this assessment for too long as, again, it can be rather uncomfortable for the patient.

The causes of carpal tunnel syndrome are wide and varied and can include pregnancy, arthritis, wrist trauma, hypothyroidism, or diabetes mellitus. These conditions need to be carefully managed, but it may be necessary for immediate treatment of the carpal tunnel syndrome.

Investigation of carpal tunnel syndrome can be undertaken by a comprehensive history and examination of the wrist joint, including Tinel's test and Phalen's sign. If arthritis is considered as a cause of the carpal tunnel syndrome, an X-ray may be required to confirm this. Some practitioners like to undertake electromyograms or nerve-conduction studies (Mayoclinic, 2015), but a lot can be gained from the history and examination.

Treatment of carpal tunnel syndrome can be done by providing a wrist splint for the patient, administering anti-inflammatory drugs, or local steroid injections. Local steroid injections tend not to last very long, and it may be necessary to undertake surgical decompression of the carpal tunnel. This can be done by an open procedure or by endoscopic means.

2.2.9 Medial Cutaneous Nerve of Arm

The medial cutaneous nerve of arm, also referred to as the medial brachial cutaneous nerve, is the smallest branch of the brachial plexus arising from the side branches of the medial cord from the eighth cervical and first thoracic nerves.

It courses through the axilla initially posterior and then runs medial to the axillary and brachial veins. It communicates with the intercostobrachial nerve.

It passes inferiorly and medially to the brachial artery to the middle region of the arm and passes through the deep fascia. Race and Saldana (1991) had shown that the course of the medial cutaneous nerve of arm was predictable relative to the basilica vein and the medial epicondyle.

The medial cutaneous nerve of the arm innervates the skin of the medial side of the arm, to the medial condyle of the humerus, and the olecranon of the ulna.

2.2.9.1 Clinical Examination

When undertaking any clinical history or examination, you should always do the following, and follow a logical and systematic format:

1. Introduce yourself to the patient.
2. Advise them of what position you hold, for example, student, specialty grade, consultant, etc.
3. Explore your reason for consulting with them, or to find out why they have presented to you.
4. Always take a thorough and detailed history that will be guided by the presenting signs and symptoms.
5. When examining the patient, always tell them what you will ask them to do, or what region of the body you will be examining, with specific instructions and ensure they give consent.

A detailed examination and history taking should be completed as described in Chapter 1.

Testing of the medial cutaneous nerve of arm can be assessed when examining the dermatome distribution, as there may be overlap in the sensory innervation of the limbs. Again, this is described within the International Standards for Neurological Classification of Spinal Cord Injury (ISNCSCI, 2015).

2.2.10 Medial Cutaneous Nerve of Forearm

The medial cutaneous nerve of forearm (Figure 2.5) arises from the side branches of the medial cord from the eighth cervical and first thoracic nerves. Initially it courses along with the ulnar nerve and pierces the deep fascia along with the basilica vein to enter the subcutaneous tissue. It has anterior and posterior branches and it innervates the medial side of the forearm as far as the wrist joint.

The anterior branch is the larger of these two branches and typically passes anterior to the median cubital vein. It then courses on the anterior aspect of the ulnar aspect of the forearm and can communicate with the palmar cutaneous branch from the ulnar nerve.

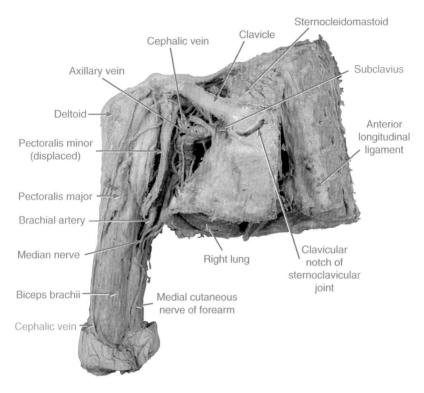

Fig. 2.5. The position of the medial cutaneous nerve of forearm.

The posterior branch passes on the medial aspect of the basilica vein anterior to the medial epicondyle, toward the posterior aspect of the forearm. It communicates with the cutaneous branches of the radial and ulnar nerves.

2.2.10.1 Clinical Examination
When undertaking any clinical history or examination, you should always do the following, and follow a logical and systematic format:

1. Introduce yourself to the patient.
2. Advise them of what position you hold, for example, student, specialty grade, consultant, etc.
3. Explore your reason for consulting with them, or to find out why they have presented to you.
4. Always take a thorough and detailed history that will be guided by the presenting signs and symptoms.

5. When examining the patient, always tell them what you will ask them to do, or what region of the body you will be examining, with specific instructions and ensure they give consent.

A detailed examination and history taking should be completed as described in Chapter 1.

Testing of the medial cutaneous nerve of forearm can be assessed when examining the dermatome distribution, as there may be overlap in the sensory innervation of the limbs. This is described within the International Standards for Neurological Classification of Spinal Cord Injury (ISNCSCI, 2015).

2.2.11 Ulnar Nerve

The ulnar nerve originates form the seventh and eight cervical nerves, as well as the first thoracic nerve, that is, C7–T1. It arises from the medial cord and typically has a root that would come from the lateral root from either the median nerve, or the lateral cord. At its origin, the ulnar nerve is found sandwiched between the axillary vein and artery, anterior to teres major.

The ulnar nerve then passes medial to the axillary artery and then continues its path down the arm lying medial to the brachial artery. At approximately mid-arm level, the ulnar nerve then pierces the medial intermuscular septum and then descends with the superior ulnar collateral artery and the ulnar collateral nerve. At that site, it is found on the medial head of the triceps muscle. It then passes to the posterior aspect of the medial epicondyle, often giving a small branch to the elbow joint. Then, the ulnar nerve passes into the forearm between the two heads of the flexor carpi ulnaris.

In the forearm, the ulnar nerve gives off a *dorsal branch*. This branch passes downward and posteriorly, between the ulna and the flexor carpi ulnaris toward the medial side of the back of hand. After giving fine branches to the back of the hand (dorsal surface), the ulnar nerve then divides into three dorsal digital nerves. These supply the little finger, ring finger, and the middle finger (sometimes to the distal interphalangeal joint) to its medial half on the dorsal aspect. The most lateral of the dorsal digital nerves communicate with the superficial branch of the radial nerve. Occasionally, a fourth digital nerve (dorsal) may be present and can extend to the index finger.

The dorsal digital branches of the radial and also the ulnar nerves do not extend all the way to the tips of the fingers. In the first and the fifth fingers, the branches do extend as far as the nail beds, but in the intermediate three fingers, the innervation typically passes to the middle phalanx, or even only to the proximal phalanx. The palmar digital branches arising from the median and the ulnar nerves then complete the rest of the digits toward the nail beds.

In the inferior portion of the forearm, the ulnar nerve can give off a *palmar branch*. This branch is a variable branch and it crosses the flexor retinaculum and innervates the skin of the medial side of the palm. The ulnar nerve then continues its journey inferiorly to enter the hand in front of the carpal tunnel. Therefore, it is outside of this tunnel, is lateral to the pisiform, and passes in front of the flexor retinaculum. The ulnar nerve is found medial to the ulnar artery, and sometimes the superficial portion of the flexor retinaculum covers these structures. Occasionally, the palmaris brevis and the pisohamate ligament may well cover the ulnar nerve and artery.

The *superficial branch of the ulnar nerve* gives a fine branch to the palmaris brevis. The superficial branch then divides into palmar digital nerves that pass to the medial side of the little finger and adjacent sides of the little and ring fingers. These branches innervate both skin and articular structures, specifically to the metacarpophalangeal and interphalangeal joints, as do those of the median nerve, and these communicate with the median nerve. Sometimes, the ulnar nerve can also supply the adjacent sides of the ring and middle fingers.

The *deep branch of the ulnar nerve* goes between the flexor digiti minimi and the abductor digiti minimi, and supplying them as it does so. The deep branch then passes through a fibrous arch in the proximal end of the opponens digiti minimi, innervating it too. This branch will then wind round the hook of the hamate and passes laterally alongside the deep palmar arch, under the cover of a fat pad behind the flexor tendons. During this journey, the deep branch innervates the adductor pollicis, third and fourth lumbricals, all of the interossei and typically the flexor pollicis brevis. Due to its extensive innervation, it has been

Table 2.11 The Branches of the Ulnar Nerve, and the Muscles, Skin, and Joints That it Innervates

Ulnar Nerve Branches		
Dorsal branch	Twigs to back of hand	
	Dorsal digital nerves	Little finger, ring finger, and middle fingers dorsally
Palmar branch	Medial side of palm	
Superficial branch	Palmaris brevis	
	Palmar digital nerves	Index finger and medial half of ring finger
		Metacarpophalangeal and interphalangeal joints
Deep branch	Flexor digiti minimi Abductor digiti minimi Adductor pollicis Third and fourth lumbricals All interossei Flexor pollicis brevis	

referred to as the *"musician's nerve"* as it innervates the muscles responsible for fine movements of the fingers. In contrast, the median nerve has been referred to as the *"laborer's nerve"* as it innervates muscles typically used in those professions, not involved in fine finger movements.

In summary, the ulnar nerve innervates the palmar aspects of the medial one and a half fingers and the dorsal aspects of the medial two and a half fingers. The median nerve, however, innervates the other fingers on the palmar surface, and the radial nerve innervates the dorsal aspect. The median nerve extends dorsally on the distal phalanx of the thumb and distal two phalanges of the first two and a one-half fingers. This is summarized in Table 2.11.

The muscles innervated by the ulnar nerve are described in Table 2.12. This provides a summary of the origin, insertion, innervation, and the function of each of these muscles.

2.2.11.1 Clinical Examination
When undertaking any clinical history or examination, you should always do the following, and follow a logical and systematic format:

1. Introduce yourself to the patient.
2. Advise them of what position you hold, for example, student, specialty grade, consultant, etc.

Table 2.12 The Muscular Innervation of the Ulnar Nerve, and Where Each Muscle Originates From, Where it Inserts, and Also Describes the Function of Each Muscle

Muscle	Origin	Insertion	Innervation	Function
Palmaris brevis	Palmar aponeurosis Flexor retinaculum	Skin on ulnar side of the palm	Superficial branch of the ulnar nerve	Tension of the skin of the palm Deepens the hollow of the palm
Flexor digiti minimi (brevis)	Hook of hamate	Medial side of base of proximal phalanx of the fifth digit	Deep branch of the ulnar nerve	Flexion of the fifth (little) digit
Abductor digiti minimi	Pisiform	Medial side of base of proximal phalanx of the fifth digit	Deep branch of the ulnar nerve	Abduction of fifth (little) digit
Adductor pollicis	Oblique head – anterior aspect of the base of the second metacarpal, capitate and the trapezoid Transverse head – third metacarpal	Medial sesamoid bone Medial side of the base of the proximal phalanx of thumb Extensor expansion	Deep branch of the ulnar nerve	Adduction of the thumb Aids opposition
Lumbricals (3 and 4)	Flexor digitorum profundus	Extensor expansion	Deep branch of the ulnar nerve (N.B. The first and second lumbricals are innervated by the median nerve)	Flexion of the metacarpophalan-geal joints Extension of the interphalangeal joints
Flexor pollicis brevis	Trapezium Flexor retinaculum	Proximal phalanx of the thumb	Deep branch of the ulnar nerve for its medial side; recurrent branch of the median nerve	Flexion of the thumb at the first metacarpophalan-geal joint

3. Explore your reason for consulting with them, or to find out why they have presented to you.
4. Always take a thorough and detailed history that will be guided by the presenting signs and symptoms.
5. When examining the patient, always tell them what you will ask them to do, or what region of the body you will be examining, with specific instructions and ensure they give consent.

A detailed examination and history taking should be completed as described in Chapter 1.

Table 2.13 summarizes the nerve roots, muscle supplied, and relevant clinical test to ask the patient to perform in assessing the ulnar nerve

Table 2.13 The Origin of Each of the Vertebral Levels of the Ulnar Nerve, the Muscles They Supply

Nerve Root	Muscle	Ask Patient to…
C7–C8	Flexor carpi ulnaris	Abduct little finger and identify the tendon when all the fingers are fully extended
C8–T1	Flexor digitorum profundus (digits 3 and 4)	Fix the middle phalanx of the little finger and resist extension to the distal phalanx
C8–T1	Dorsal interossei	Abduction of the fingers
C8–T1	Palmar interossei	Adduction of the fingers
C8–T1	Adductor pollicis	Adduction of the thumb
C8–T1	Abductor digiti minimi	Abduction of the little finger
C8–T1	Opponens digiti minimi	With all the fingers extended, bring the little finger over the other fingers

The table also states what to ask the patient to do when assessing each muscle and therefore approximate nerve root of the ulnar nerve.

as well as the assessment of the muscles that the ulnar nerve supplies. Sensation should also be assessed in the distribution of the ulnar nerve, that is, the index finger and medial half of ring finger. Again, this should be according to the International Standards for Neurological Classification of Spinal Cord Injury (ISNCSCI, 2015) as previously described in Chapter 1.

2.2.11.2 Clinical Applications

1. *Ulnar nerve injury*: Injury to the ulnar nerve tends to happen in one of four key areas:
   a. Behind the medial epicondyle of the humerus
   b. Cubital tunnel, that is, a tunnel bordered by the medial epicondyle of the humerus, olecranon of the ulna and the tendinous arch formed by the ulnar and humeral heads of the flexor carpi radialis
   c. Wrist joint
   d. Hand

   Typically, injury to the ulnar nerve happens as it runs behind the medial epicondyle of the humerus, and can occur with a fracture at that site.

   Injury to the ulnar nerve will present with numbness and paresthesia in the little finger, the medial half of the ring finger, and the medial side of the palm. As well as the sensory deficit, there can be extensive motor loss to the hand. This will result in reduced wrist adduction when trying to flex the wrist, and the hand will be drawn over to the radial side due to the unopposed pull of the flexor carpi radialis that is innervated by the median nerve. This is due to a lack of innervation of the flexor carpi ulnaris by the ulnar nerve. The patient will also have difficulty making a fist because there is a lack of opposition, and the metacarpophalangeal joints will become hyperextended and unable to flex the fourth and fifth digits at the distal interphalangeal joints. In addition to this, the patient will be unable to extend the interphalangeal joints when trying to straighten the digits. The characteristic appearance of the patient's hand will result in a *"claw hand"* due to wasting of the interossei. Therefore, the "claw hand" appearance is because of the unopposed action of the extensors and also the flexor digitorum profundus.

2. *Cubital tunnel syndrome*: Tunnel is created by the ulnar and humeral heads of the flexor carpi ulnaris. If the ulnar nerve is compressed at that point, it is referred to as cubital tunnel syndrome. This syndrome can be caused by sleeping with the arms behind the neck, and the elbow joint flexed, or pressing the elbows when in the typing position for office workers, or indeed from strenuous bench-press exercises. The patient will complain of pain and numbness (paresthesia) along the sensory distribution of the ulnar nerve, that is, the medial one and a half fingers as well as the medial portion of the palm. It may be treated by conservative measures like removing or alleviating the exacerbating factor, and may require surgery.

3. *Guyon's tunnel syndrome*: Guyon's canal or tunnel, or the ulnar canal or tunnel is a canal formed at the wrist joint by the following structures that allow the ulnar nerve and artery a route to pass into the hand:
   a. *Superior boundary*: Superficial palmar carpal ligament
   b. *Floor*: Hypothenar muscles and the deep flexor retinaculum
   c. *Medial boundary*: Pisiform and pisohamate ligament
   d. *Lateral boundary*: Hook of the hamate
   Guyon's tunnel syndrome results in the patient complaining of reduced sensation or numbness, paresthesia of the medial one and a half fingers, and weakness of the intrinsic muscles of the hand. Clawing of the hand may be present with hyperextension

of the metacarpophalangeal joints with flexion at the proximal interphalangeal joint affecting the fourth and fifth digits. Flexion is not affected and radial deviation is not present.

Shea and McClain (1969) divided pathologies affecting the ulnar nerve into three types. Type I involved pathology affecting the ulnar nerve proximal to or within Guyon's canal. This would exhibit both motor and sensory abnormalities. Type II affected only the deep branch of the ulnar nerve and resulted in weakness in muscles innervated by that branch, and could spare the hypothenar muscles. Type III involves compression of the ulnar nerve at the end of Guyon's canal, resulting in only a sensory deficit and no motor-related problems.

Typical problems to affect the ulnar nerve and result in Guyon's tunnel syndrome arise from a ganglion, trauma, musculotendinous arch, or disease of the ulnar artery, and surgery may be offered to these patients in alleviating the pain, numbness, and muscle weakness (Aguiar et al., 2001).

2.2.12 Upper Subscapular Nerve

A variable number of subscapular branches arise from the posterior cord of the brachial plexus, and a twig is often given to the shoulder joint as well. The upper subscapular nerve, or nerves, arise from the fifth cervical nerve and innervates subscapularis.

2.2.13 Lower Subscapular Nerve

The lower subscapular nerve, or nerves, arise from the fifth and sixth cervical nerves and not only innervate the subscapularis but also the teres major.

Subscapularis forms part of the posterior wall of the axilla. It arises from almost the whole of the subscapular fossa. Its tendon of insertion passes in front of the capsule of the shoulder joint, to which it is adherent, and is attached to the lesser tubercle of the humerus and its crest. The subscapularis is a strong medial rotator of the arm and helps to hold the head of the humerus in the glenoid cavity.

Tears of the subscapularis muscle and related tendon may have been underreported, and are more obvious nowadays due to the advancement of arthroscopic techniques (Garavaglia et al., 2011). Dependent on the cause and severity of the tear, surgery may be a required option due to the limitation of shoulder movement. Clinical examination of the shoulder joint should follow that described in Section 2.2.3. Internal (medial) rotation of the shoulder joint will be affected most with this condition.

Teres major arises from the dorsal surface near the inferior angle, and is inserted into the crest of the lesser tubercle below the insertion of the subscapularis. The lower edges of the tendons of the latissimus dorsi and teres major are commonly fused near their insertions; the tendons are otherwise separated by a bursa. Another bursa intervenes between the tendon of the teres major and the humerus. The teres major, with latissimus dorsi and subscapularis, form the posterior wall of the axilla.

The teres major acts with latissimus dorsi in adduction of the arm, and any help in medial rotation.

2.2.14 Thoracodorsal Nerve

The thoracodorsal nerve typically arises from the seventh and eighth cervical nerves (i.e., C7–C8), though can receive contribution from the sixth cervical nerve (C6). This nerve is a branch of the posterior cord of the brachial plexus. Through its journey in the axilla, it is approximately 2–3 mm in size. The thoracodorsal nerve passes under then posterior to the axillary vein and first appears in the axilla close to the lateral thoracic vein. The thoracodorsal nerve then changes its course inferiorly and laterally and amalgamates with the subscapular vessels thus forming the thoracodorsal neurovascular bundle. Zin et al. stated that a landmark that can be seen when identifying the thoracodorsal nerve is when a triangle is formed by the thoracodorsal nerve medially, the suprascapular vessels laterally, and the axillary vein superiorly. The thoracodorsal nerve innervates the latissimus dorsi, as described in Table 2.14. Latissimus dorsi is a large triangular muscle found superficially, apart from the superior part that is covered by the trapezius. Near its lower attachment, it forms the posterior boundary of the lumbar triangle. Latissimus dorsi, along with the teres major, forms the posterior axillary fold and thus contributes to the posterior wall of the axilla.

Table 2.14 The Origin, Insertion, Innervation, and Function of the Latissimus Dorsi Muscle

Muscle	Origin	Insertion	Innervation	Function
Latissimus dorsi	Spinous processes of the lower six thoracic vertebrae Spinous processes of the lumbar and sacral vertebrae (indirectly) Iliac crest	Floor of the intertubercular groove	Thoracodorsal nerve	Adductor and extensor of arm Rotates arm medially Accessory muscle of respiration

The latissimus dorsi arises from several components (Table 2.14). It arises from the spinous processes of the lower six thoracic vertebrae, indirectly from the spinous processes of the lumbar and sacral vertebrae via its attachment to the posterior layer of the thoracolumbar fascia, and also from the iliac crest. As the muscle passes to its insertion, it receives slips from the outer surface of the lower three or four ribs, and typically from the inferior angle of the scapula. The latissimus dorsi then spirals around the lower edge of the teres major and inserts into the floor of the intertubercular groove. The lower edges of the tendons of the latissimus dorsi and teres major are commonly fused near their insertions.

The latissimus dorsi is a powerful adductor and extensor. It also can rotate the arm medially, and is responsible for the down stroke of the arm in swimming due to these actions. It can also be typically used in climbing, hammering, rowing, and supporting the weight of the body on the hands. Its scapular attachment can also assist in keeping the inferior angle of the scapula against the chest wall. Through its attachments to the ribs, it can also act as an accessory muscle of respiration.

2.2.14.1 Clinical Examination
When undertaking any clinical history or examination, you should always do the following, and follow a logical and systematic format:

1. Introduce yourself to the patient.
2. Advise them of what position you hold, for example, student, specialty grade, consultant, etc.
3. Explore your reason for consulting with them, or to find out why they have presented to you.
4. Always take a thorough and detailed history that will be guided by the presenting signs and symptoms.

5. When examining the patient, always tell them what you will ask them to do, or what region of the body you will be examining, with specific instructions and ensure they give consent.

A detailed examination and history taking should be completed as described in Chapter 1.

There are two main ways that the function of latissimus dorsi can be assessed, and therefore the integrity of the thoracodorsal nerve (C6–C8). With both techniques, both sides should be compared to assess for symmetry and to note any obvious difference, for example, muscle wasting, fasciculations, etc.

1. The patient can be asked to adduct the arm with the forearm flexed into 90° with the palm facing forward. The power of this muscle can then be assessed by the examiner.
2. With the examiner's hands placed over the back of the patient or over the posterior fold of the axilla with the upper limb abducted, ask the patient to cough. The contraction of the muscle will be easily felt. It is essential to compare both sides.

2.2.14.2 Clinical Applications

As the latissimus dorsi has many substitutes for its actions, related clinical defects are actually rather difficult to identify.

As the latissimus dorsi is such a large muscle, smaller segments with its related neurovasculature can therefore be used in facial reanimation procedures. Due to its long nerve that innervates this muscle, the thoracodorsal nerve, this therefore avoids the need for a two-stage facial reanimation procedure.

Through cadaveric dissection, Ferguson et al. (2011) had shown that the thoracodorsal pedicle was indeed long at between 10 and 14 cm. They had also shown that small muscle segments could also be raised measuring 8–10 cm × 2–3 cm and suggested that with its long neurovascular pedicle, it could then be used to do cross-face nerve grafting for those cases requiring only a single facial reanimation procedure. Ferguson et al. (2011) had shown through four complex cases of facial paralysis that the use of segmental parts of the latissimus dorsi could in fact be used for single-stage facial reanimation procedures, for both static and dynamic reanimation.

2.2.15 Axillary Nerve

The axillary nerve arises from the fifth and the sixth cervical nerves, that is, C5–C6. The axillary nerve is a branch from the *posterior cord*. It is found anterior to the subscapularis, posterior to the brachial artery, and lateral to the radial nerve. At the inferior end of the subscapularis the axillary nerve then runs posterior, close to the joint capsule, passing through the quadrangular space with the posterior circumflex humeral artery, sandwiched between the lateral and long heads of the triceps muscle. The axillary nerve is found inferior to the capsule of the shoulder joint, and it sends a small branch to this joint. The axillary nerve then winds medial to the surgical neck of the humerus, and is typically in contact with this part of the bone. The axillary nerve then usually divides into two branches at this point under the cover of the deltoid muscle. The *anterior branch of the axillary nerve* winds round the humerus deep to the deltoid muscle, and also innervates the muscle at this point (Table 2.15). The *posterior branch of the axillary nerve* innervates the teres minor (Table 2.15) and also the deltoid. The posterior branch then winds round the deltoid muscle and goes to innervate an area of skin on the back of the arm as the upper lateral brachial cutaneous nerve, or the superior lateral cutaneous nerve of arm. This nerve innervates specifically skin over the inferior two-thirds of the deltoid muscle at its posterior aspect. Superior to this, the skin of the shoulder is innervated by the supraclavicular nerves. The level of the axillary nerve can also be thought of as by a horizontal plane through the middle of the deltoid muscle.

Table 2.15 The Muscles Which are Innervated by the Axillary Nerve in Terms of Their Origin, Insertion, and Actions

Muscle	Origin	Insertion	Innervation	Function
Deltoid	Front of superior surface of the lateral third of the clavicle, adjoining acromion, spine of the scapula	Deltoid tuberosity of the humerus	Axillary nerve	Acromial part – powerful abductor of arm Spinous part – extension of arm and lateral rotation Clavicular part – flexion of arm and medial rotation
Teres minor	Lateral margin of the infraspinous fossa	Capsule of the shoulder joint Lowest facet on the greater tubercle of the humerus	Axillary nerve	Lateral rotation of the arm Maintains the head of the humerus in place during abduction

2.2.15.1 Clinical Examination

When undertaking any clinical history or examination, you should always do the following, and follow a logical and systematic format:

1. Introduce yourself to the patient.
2. Advise them of what position you hold, for example, student, specialty grade, consultant, etc.
3. Explore your reason for consulting with them, or to find out why they have presented to you.
4. Always take a thorough and detailed history that will be guided by the presenting signs and symptoms.
5. When examining the patient, always tell them what you will ask them to do, or what region of the body you will be examining, with specific instructions and ensure they give consent.

A detailed examination and history taking should be completed as described in Chapter 1.

Specific to the axillary nerve, the following should be examined:

1. *Assessment of the deltoid*: This can be undertaken by asking the patient to fully abduct their arm. Symmetry over the deltoid muscle should be noted and the left and right sides compared.
2. *Assessment of the teres minor*: The function and power of the teres minor can be assessed by the *Hornblower's test*. The upper limb should be abducted by the examiner out to 90° in the plane of the scapula, then flex the elbow. The examiner should place medial pressure at the elbow and hold the hand in position while asking the patient to perform external (lateral) rotation against resistance by the examiner. A positive Hornblower's test would result in pain and/or dysfunction during this procedure. The other way to do this is by asking the patient to bring their hands toward their mouth, but without abducting the shoulder joint. Therefore, this test assesses the posterior part of the rotator cuff muscles. The teres minor is required to bring the patient's hands toward their mouth and stabilize the lateral rotation of the shoulder joint.
3. *Sensory assessment*: The axillary nerve also provides innervation to skin over the inferior two-thirds of the deltoid muscle at its posterior aspect. This is referred to as the regimental badge area. Injury to

the axillary nerve will result in reduced or absent sensation over this anatomical territory.

2.2.15.2 Clinical Applications
2.2.15.2.1 Fractures of the Surgical Neck of the Humerus
The axillary nerve can be damaged due to a fracture of the surgical neck of the humerus as the nerve winds around this point, just inferior to the head of the humerus. The axillary nerve may also be damaged by fractures to the surgical neck of the humerus, shoulder (glenohumeral) dislocation, or from crutches pressing in the axilla through improper use or provision.

On examination, the patient would have a flattened deltoid and also there may be pitting over the acromion. There will be sensory loss as previously described.

2.2.15.2.2 Intramuscular Injection to Deltoid
The deltoid may be used as a muscle to give intramuscular injections too. The clinician should be aware of the axillary nerve running under the deltoid at the surgical neck of the humerus. This should therefore prevent injury to this nerve.

2.2.15.2.3 Axillary Brachial Plexus Block
The axillary brachial plexus block is typically performed for hand and forearm surgery, and should be undertaken using ultrasound guidance. This technique of anesthetizing the brachial plexus is considered superior compared to supraclavicular or interscalene blocks. The anesthesia extends from the mid-arm level down to the hand. Although it can be referred to as the axillary brachial plexus block, this is only due to the access to the brachial plexus via the axilla and does NOT anesthetize the axillary nerve due to its origin form the posterior cord high up within the axilla (NYSORA, 2015).

2.2.16 Radial Nerve
The radial nerve arises from the posterior cord of the brachial plexus. As the largest branch of the brachial plexus, it courses behind the third part of the axillary artery and also the superior part of the brachial artery. As it passes behind the brachial artery it soon dips forward with the profunda brachii artery. It passes anterior to the tendons of teres

major and latissimus dorsi and runs in front of the subscapularis. With
the profunda brachii artery, and later on, its radial collateral branch,
the radial nerve then passes around the humerus obliquely. It then
passes between the lateral and medial heads of the triceps brachii, and
then within a shallow groove deep to the lateral head of the triceps.

Once the radial nerve reaches the outer aspect of the humerus, it will
then pierce the intermuscular septum (lateral), and then pass to the anterior compartment of the arm. It then passes sandwiched between the brachialis medially and the extensor carpi radialis inferorlaterally and the
brachioradialis superolaterally. At or below the level of the lateral epicondyle, the radial nerve then divides into its superficial and deep branches.

In terms of surface anatomy, the radial nerve extends from the medial
margin of the biceps, opposite the posterior axillary fold, obliquely
across the back of the arm. It pierces the intermuscular septum at the
superior point of trisection of a line between the deltoid insertion and
the lateral epicondyle and then descends to the front of the lateral epicondyle. The radial nerve can be palpated in thin individuals as it winds
around the humerus. Typically, it can be felt in these individuals about
1–2 cm inferior to the deltoid insertion, and also in the interval between
the brachialis and the brachioradialis.

In the arm the radial nerve can be found if a line is imaginarily drawn
from the start of the brachial artery across the elevations from the lateral and long heads of the triceps brachii muscle to the junction of the
upper and middle thirds if a line is drawn from the deltoid tuberosity to
the lateral epicondyle. Again, the radial nerve can also be found approximately 10 mm lateral to the tendon of the biceps brachii muscle.

The radial nerve has the following branches: *cutaneous, articular,
muscular, and superifical and deep terminal branches.*

1. *Cutaneous branches*

   The cutaneous nerves of the radial nerve are the lower lateral
   cutaneous nerve of the arm and the posterior cutaneous nerve of
   the arm.

   The lower lateral (brachial) cutaneous nerve of the arm pierces the
   lateral head of the triceps brachii muscle inferior to the insertion

of the deltoid muscle. The lower lateral cutaneous nerve of the arm then courses anterior to the elbow, closely related to the cephalic vein and innervates the skin of the lower portion of the inferior half of the arm.

The posterior (brachial) cutaneous nerve of the arm is smaller and originates in the axilla and courses to the inner aspect of the arm innervating the skin of the dorsal aspect of the arm almost to the olecranon. It also passes behind and communicates with the intercostobrachial nerve. The intercostobrachial nerve is a cutaneous branch of the intercostal nerve.

2. *Articular branches*
   The articular branches of the radial nerve innervate the elbow joint.
3. *Muscular branches*
   The muscular branches of the radial nerve innervate the extensor carpi radialis longus, brachialis, brachioradialis, anconeus, and triceps. These are grouped as lateral, medial, and posterior.
   The *lateral muscular branches* originate form the nerve as they are in front of the lateral intermuscular septum. The lateral muscular branches innervate the extensor carpi radialis longus, brachioradialis, and the brachialis.
   The *medial muscular branches* originate from the radial nerve on the inner aspect of the arm and innervate the long and medial heads of the triceps. The branch that innervates the medial head of the triceps is a long thin branch that is positioned very close to the ulnar nerve down to the distal one-third of the arm and can also be termed the *ulnar collateral nerve.*
   The posterior muscular branch is larger and arises from the radial nerve as it is positioned within the groove. It innervates the anconeus, as well as the lateral and medial heads of the triceps muscle. The branch to the anconeus is a long nerve and this branch is within the substance of the medial head of the triceps muscle, and innervates it too. This branch is also accompanied by the profunda brachii artery; it goes posterior to the elbow joint and then terminates in the anconeus.
4. *Superficial branch*
   The superficial or superficial terminal branch passes inferior and anterior to the lateral epicondyle and along the outer aspect of the

upper two-thirds of the forearm. First it lies on supinator, lateral to the radial artery and posterior to the brachioradialis. It then is found posterior to the brachioradialis in the middle one-third of the forearm and runs close to the lateral side of the radial artery. It is then found on the pronator teres then on the flexor digitorum superficialis (radial side) and then positioned on the flexor pollicis longus. Further, it moves away from the radial artery approximately 7 cm above the wrist. The superficial terminal branch will then pass under the tendon of the brachioradialis and swings around the lateral aspect of the radius as it passes distally and it goes through the deep fascia and terminates in four or five digital nerves.

5. *Dorsal digital nerves*

   The dorsal digital nerves are small and there can be either four or five of these. The dorsal digital nerves frequently join with the lateral and posterior cutaneous nerves (mentioned earlier). Note that the digital nerves innervating the thumb only pass as far as the root of the nail. The dorsal digital nerves innervating the middle finger only reach the middle phalanx and those to the middle and ring finger do not pass any further than the proximal interphalangeal joints. The rest of the distal aspect of the skin on the dorsal aspect of the digits are innervated by the median and ulnar nerves' palmar digital branches. Table 2.16 summarizes what the branches supply.

6. *Deep terminal branch of the radial nerve*

   The deep terminal branch of the radial nerve is also referred to as the posterior interosseous nerve. The deep branch of the radial nerve is muscular and articular in its distribution. It originates from under the brachioradialis and passes laterally around the radius, sandwiched between the superficial and deep layers of the supinator that it also innervates. It typically will contact the bare surface of the radius and

Table 2.16 The Innervation Territories of the Dorsal Digital Nerves	
Dorsal Digital Nerve	Innervation
1st	Skin of radial aspect of the thumb and adjacent part of the thenar eminence
2nd	Medial aspect of the thumb
3rd	Lateral aspect of the index finger
4th	Adjoining side of index and middle finger
5th	Adjoining aspect of the middle and ring finger

can be susceptible to damage if there is a fracture of the radius at this site. When the deep terminal branch of the radial nerve reaches the back of the forearm, it is found between the superficial and deep extensors, giving innervation to the superficial group, and is accompanied by the posterior interosseous artery. For the rest of the course of tuso nerve, it is termed the posterior interosseous nerve, a name that has also been applied to the deep terminal branch in its entirety. In the distal part of the forearm, it passes onto the interosseous membrane by going deep to the extensor pollicis longus. It then is found with the anterior interosseous artery, in the groove for the extensor digitorum on the back of the radius. It terminates on the back of the carpus in an enlargement from which twigs are sent to the wrist joint and the intercarpal joints.

As the deep terminal branch of the radial nerve passes through the forearm, they supply the supinator (and frequently the extensor carpi radialis brevis), the extensor digitorum, extensor digiti minimi and the extensor carpi ulnaris. The posterior interosseous nerve innervates the abductor pollicis longus, extensor pollicis brevis, extensor pollicis longus, as well as the extensor indicis.

A summary of the branches of the radial nerve branches is given in Table 2.17. The muscles innervated by the radial nerve are summarized in terms of their origin, insertion, actions, and nerve supply in Table 2.18.

2.2.16.1 Clinical Examination

When undertaking any clinical history or examination, you should always do the following, and follow a logical and systematic format:

1. Introduce yourself to the patient.
2. Advise them of what position you hold, for example, student, specialty grade, consultant, etc.
3. Explore your reason for consulting with them, or to find out why they have presented to you.
4. Always take a thorough and detailed history that will be guided by the presenting signs and symptoms.
5. When examining the patient, always tell them what you will ask them to do, or what region of the body you will be examining, with specific instructions and ensure they give consent.

Table 2.17 The Branches of the Radial Nerve and the Motor, Sensory, or Articular Innervations of Each Branch

Branch	Innervation
Cutaneous	
Lower lateral	Skin of lower portion of the inferior half of the arm
Posterior	Dorsal aspect of arm (to olecranon)
Articular	Elbow joint
Muscular	
Lateral	Brachioradialis Extensor carpi radialis longus Brachialis
Medial	Long and medial heads of the triceps
Posterior	Lateral and medial heads of the triceps muscle Anconeus
Superficial	Terminates in the dorsal digital nerves
Dorsal digital nerves	
1st	Skin of radial aspect of the thumb and adjacent part of the thenar eminence
2nd	Medial aspect of the thumb
3rd	Lateral aspect of the index finger
4th	Adjoining side of index and middle finger
5th	Adjoining aspect of the middle and ring finger
Deep terminal branch	Supinator Extensor carpi radialis brevis Extensor digitorum Extensor digiti minimi Extensor carpi ulnaris Abductor pollicis longus Extensor pollicis brevis Extensor pollicis longus Extensor indicis

A detailed examination and history taking should be completed as described in Chapter 1.

Table 2.19 will highlight what nerve roots of the radial nerve have to be tested, what muscle it supplies, a brief description of the normal function, and how to test for it clinically.

2.2.16.2 Clinical Applications

Damage to the radial nerve may happen at several points along its course. Typically, the radial nerve may be damaged by trauma or entrapment of the nerve.

Table 2.18 Muscles Supplied by Specific Branches of the Radial Nerve (Including Vertebral Level Origin), its Origin, Insertion, and Functions

Muscle	Origin	Insertion	Actions	Innervation
Brachialis	Distal two-thirds of the anteromedial and anterolateral surface of the humerus	Capsule of the elbow joint Coronoid process Tuberosity of the ulna	Flexion of the forearm	Musculocutaneous nerve (C5–C6) and a small contribution from the radial nerve (C7) to its lateral aspect
Brachioradialis	Upper part of the lateral supracondylar ridge of the humerus	Lateral surface of the radius (just superior to the styloid process)	Flexion of the forearm when mid-pronated	Radial nerve (C5–C6)
Triceps	Long head – infraglenoid tubercle of the scapula Lateral head – posterior surface of the humerus above grove for radial nerve Medial head – posterior surface of the humerus below grove for radial nerve	Posterior portion of the upper aspect of the olecranon Fascia of the forearm via the tricipital aponeurosis	Extension of the forearm Lateral and long heads recruited when resistance occurs Long head can also hold the humerus in place during its actions	Radial nerve (C6–C8)
Anconeus	Lateral epicondyle of humerus	Olecranon and posterior surface of ulna	Assists triceps in extension of the forearm Active during pronation	Radial nerve (C6–C8)
Extensor carpi radialis longus	Lower portion of the lateral supracondylar ridge	Back of the base of the 2nd metacarpal	Extension of the hand Abduction of the hand at the wrist joint	Radial nerve (C6–C7)
Supinator	Superficial part – lateral epicondyle of the humerus Deep part – supinator fossa and crest Oblique line of ulna	Superficial part – radius Deep part – upper third of shaft of the radius	Supination of the forearm	Posterior interosseous nerve (C6–C7)
Extensor carpi radialis brevis	Lateral epicondyle of humerus	Back of the bases of the 2nd and 3rd metacarpals	Extension of the hand Abduction of the hand at the wrist joint	Deep branch of radial nerve (C7–C8)
Extensor digitorum	Lateral epicondyle of humerus	Extensor expansion and the middle and distal phalanges of fingers 2–5	Extends the proximal phalanges on the metacarpals	Posterior interosseous nerve (C7–C8)
Extensor digiti minimi	Lateral epicondyle of humerus	Extensor aponeurosis of the little finger	Extension of the proximal phalanx of the little finger	Posterior interosseous nerve (C7–C8)

Table 2.18 Muscles Supplied by Specific Branches of the Radial Nerve (Including Vertebral Level Origin), its Origin, Insertion, and Functions *(cont.)*

Muscle	Origin	Insertion	Actions	Innervation
Extensor carpi ulnaris	Lateral epicondyle of humerus	Tubercle on medial side of base of 5th metacarpal	Extends hand and acts with the radial extensors. Pure adduction when acting with flexor carpi ulnaris	Posterior interosseous nerve (C7–C8)
Abductor pollicis longus	Interosseous membrane Posterior surface of the radius and ulna	Lateral aspect of the base of the first metacarpal and (typically) the trapezium	Abduction of the first metacarpal at the carpometacarpal joint Stabilizes the first metacarpal during movement of the phalanges	Posterior interosseous nerve (C7–C8)
Extensor pollicis brevis	Posterior surface of the radius	Back of the proximal phalanx of the thumb	Extension of the thumb at metacarpophalangeal joint	Posterior interosseous nerve (C7–C8)
Extensor pollicis longus	Posterior surface of ulna (middle) Interosseous membrane	Base of distal phalanx of thumb (dorsal aspect)	Extension of distal phalanx via extension of the interphalangeal joint of the thumb	Posterior interosseous nerve (C7–C8)
Extensor indicis	Posterior surface of the ulna Interosseous membrane	Extensor expansion of the index finger	Extension of the proximal phalanx of the index finger	Posterior interosseous nerve (C7–C8)

1. *Fracture of the shaft of the humerus*

   If the humerus is fractured at mid-humerus level, the radial nerve could be injured in the radial grove (or radial sulcus) of the humerus. If there is a fracture at this site, it tends not to affect the triceps as the nerve supply to two of the three heads of the triceps muscle tends to arise more proximal to this point. However, there may be weakness of the triceps muscle and muscles in the posterior compartment of the forearm may become weakened or paralyzed. This would result in paralysis of the brachioradialis, supinator, and also the extensor muscles within the fingers and hand. This would result in the clinical presentation of *wrist-drop*. This is seen clinically with the inability to extend the fingers and wrist joint at the metacarpophalangeal joints. The "relaxed" wrist joint becomes more flexed due to the unopposed contraction of the flexor muscles of the wrist and fingers, hence the *wrist-drop*. Palsy of the radial nerve has been said to be the most common nerve lesion affecting the long bones due to fracture (Rockwood et al., 1996).

Table 2.19 The Nerve Roots That Contribute to the Radial Nerve, What Specific Musculature They Supply, the Typical Functions of Those Muscles, and the Clinical Test to Assess Function of That Nerve Root and Muscle

Nerve root	Muscle	Function	Testing
C5–C6	Brachioradialis	Flexion of the foreram when mid-pronated	Have the patient flex elbow midway between pronation and supination
C6–C7	Supinator	Supination	
C6–C7	Extensor carpi radialis longus	Extends and abducts the wrist	Extension of the patient's wrist to the side of the radius with fingers in the extended posiiton
C6–C8	Triceps brachii	Extension of the forearm at elbow joint	Extend the patient's elbow against resistance
C7–C8	Extensor digitorum	Extension of the index, middle, ring and little fingers; extension of the wrist	Ask patient to maintain their fingers in the extended position at the metacarpophalangeal joints
C7–C8	Extensor carpi ulnaris	Extension and abduction of the wrist joint	Ask patient to extend their wrist to the ulnar side
C7–C8	Abductor pollicis longus	Abduction of the first metacarpal at the carpometacarpal joint	Ask patient to abduct their thumb
C7–C8	Extensor pollicis brevis	Extension of the thumb at metacarpophalangeal joint	Ask patient to extend their thumb at the metacarpophalangeal joint
C7–C8	Extensor pollicis longus	Extension of distal phalanx via extension of the interphalangeal joint of the thumb	Ask the patient to resist flexion of the thumb at the interphalangeal joint of the first digit

2. *Radial nerve injury at the elbow*

There may be compression of the radial nerve at the elbow due to thickened fibrous tissue proximal to the radial tunnel. There may also be pronounced tissue at the upper edge of the superficial part of the supinator resulting in compression of the posterior interosseous nerve – a continuation of the deep branch of the radial nerve. This is referred to as the *arcade of Frohse*, named after the German anatomist Fritz Frohse who described it.

Radial nerve compression may also be caused by the recurrent radial artery, referred to as hypertrophic leash of Henry. This results in compression of the deep branch of the radial nerve and can present with chronic forearm pain (Loizides et al., 2011).

3. *Radial nerve injury at the wrist and hand*

The distal branches of the radial nerve may be affected by fractures of the distal radius, soft-tissue mass at the wrist or by a prominent extensor carpi radialis brevis compressing the nerve. Lesions to the superficial branches of the radial nerve could arise from a tight plaster cast at that site, handcuffs, or tight wristbands. This would result in pain and perhaps anesthesia along the distribution of the sensory branching of the radial nerve but would not result in it affecting the muscles of the hand.

REFERENCES

Aguiar, P.R., Bor-Seng-Shu, E., Gomes-Pinto, F., Almeida-Leme, R.J., Freitas, A.B.R., Martins, R.S., Nakagawa, E.S., Tedesco-Marchese, A.J., 2001. Surgical management of Guyon's canal syndrome. Arq. Neuropsiuiatar. 59 (1), 106–111.

Akuthota, V., Herring, S.A., 2009. Nerve and vascular injuries in sports medicine. Springer, Denver, USA.

Bora, Jr., F.W., Pleasure, D.E., Didizlan, N.A., 1976. A study of nerve regeneration and neuroma formation after nerve suture by various techniques. J. Hand. Surg. 1, 138–143.

Ebreheim, N.A., Lu, J., Porshinsky, B., Heck, B.E., Yeasting, R.A., 1998. Vulnerability of long thoracic nerve: an anatomical study. J. Shoulder Elbow Surg. 133, 458–461.

Ferguson, L.D., Paterson, T., Ramsay, F., Arrol, K., Dabernig, J., Shaw-Dunn, J., Molrey, S., 2011. Applied anatomy of the latissimus dorsi free flap for refinement in one-stage facial reanimation. JPRAS. 64 (11), 1417–1423.

Garavaglia, G., Ufenast, H., Taveran, E., 2011. The frequency of subscapularis tears in arthroscopic rotator cuff repairs: a retrospective study comparing magnetic resonance imaging and arthroscopic findings. Int. J. Shoulder Surg. 5 (4), 90–94.

Genetics Home Reference. http://ghr.nlm.nih.gov/condition/poland-syndrome (accessed 04.24.2015).

Gregg, J.R., Tabosky, D., Harty, M., Totke, P., Distefano, V., Das, M., 1979. Serratus anterior paralysis in the young athlete. J. Bone Joint Surg. Am. 6l (6A), 825–832.

Hamada, J., Igarashi, E., Akita, K., Mochizuki, T., 2008. A cadaveric study of the seratus anterior muscle and the long thoracic nerve. J. Shoulder Elbow Surg. 17, 790–794.

International Standards for Neurological Classification of Spinal Cord Injury (ISNCSCI). American Spinal Injury Association. http://www.asia-spinalinjury.org/elearning/ASIA_ISCOS_high.pdf (accessed 04.24.2015).

Kauppila, U., Vastamaki, M., 1996. Iatrogenic serratus anterior paralysis: long-term outcome in 26 patients. Chest 109, 3l–34.

Loizides, A., Peer, S., Ostermann, S., Henninger, B., Stampfer-Kountchev, M., Gruber, H., 2011. Unusual functional compression of the deep branch of the radial nerve by a vascular branch (leash of Henry): ultrasonographic appearance. Rofo. 183 (2), 163–166.

Machleder, H.I., 1991. Transaxillary operative management of thoracic outlet syndrome. In: Ernst, C.B., Stanley, J.C. (Eds.), Current therapy in vascular surgery. second ed. BC Decker, Inc, Philadelphia, pp. 227–230.

Mayoclinic. http://www.mayoclinic.org/diseases-conditions/carpal-tunnel-syndrome/basics/tests-diagnosis/con-20030332 (accessed 04.24.2015).

Narakas, A., 1989. Compression syndromes about the shoulder including brachial plexus. In: Szabo, R.M. (Ed.), Nerve Compression Syndromes: Diagnosis and Treatment. SLACK Incorporated, Thorofare, NJ, pp. 236–239.

NYSORA, The New York School of Regional Anesthesia. http://www.nysora.com/techniques/ultrasound-guided-techniques/upper-extremity/3017-ultrasound-guided-axillary-brachial-plexus-block.html (accessed 04.28.2015).

Pećina, M.M., Krmpotić-Nemanić, J., Markiewitz, A.D., 2001. Tunnel syndromes. Peripheral nerve compression syndromes. Crc Press, Florida, USA, pp. 71–72.

Race, C.M., Saldana, M.J., 1991. Anatomic course of the medial cutaneous nerve of the arm. J. Hand Surg. Am. 16 (1), 48–52.

Rockwood, Jr., C.A., Green, D.P., Bucholz, R.W., Heckman, J.D., 1996. Rockwood and Green's fracture in adults, fourth ed. Lippincott-Raven Publishers, Philadelphia, 1043–1045.

Roos, D.B., 1971. Experience with first rib resection for thoracic outlet syndrome. Ann. Surg. 173, 429–442.

Schultes, G., Gaggl, A., Kärcher, H., 1999. Reconstruction of accessory nerve defects with vascularized long thoracic vs. non-vascularized thoracodorsal nerve. J. Reconstr. Microsurg. 15, 265–271.

Shea, J.D., McClain, E.J., 1969. Ulnar-nerve compression syndromes at and below the wrist. J. Bone Joint Surg. 51, 1095–1103.

Sunderland, S., 1978. Nerves and Nerve Injuries, second ed. Churchill Livingstone, Edinburgh.

Wiater, J.M., Flatow, E.L., 1999. Long thoracic nerve injury. Clin. Orthop. Relat. Res. 368, 17–27.

Willison D. The Edinburgh medical and surgical journal: exhibiting a concise view of the latest and most important discoveries in medicine, surgery and pharmacy. https://books.google.co.uk/books?id=e8FOAAAAcAAJ&pg=PA129&lpg=PA129&dq=bell+and+bell+1826+external+respiratory+nerve&source=bl&ots=EiQuEVvmcC&sig=b0HTwSdgywS0Tu3Ms0IRzCSwhG-Y&hl=en&sa=X&ei=LLs3VZSVHsK6aY-PgYAI&ved=0CDYQ6AEwBA#v=onepage&q=bell%20and%20bell%201826%20external%20respiratory%20nerve&f=false (accessed 04.22.2015).

Wood, V.E., Twito, R., Verska, J.M., 1988. Thoracic outlet syndrome. The results of first rib resection in 100 patients. Orthop. Clin. North Am. 19, 131–146.

CHAPTER 3

Lower Limb Nerve Supply

3.1 OVERVIEW OF THE LOWER LIMB NERVOUS SYSTEM

The lower limb is connected by the girdle to the trunk with the upper limb, and has three individual parts to it – the *thigh, leg* and the *foot*. The pelvic girdle is formed by the two hip bones, joined anteriorly but separated posteriorly by the sacrum. The pelvic girdle and the sacrum together form a heavy-rigid ring called the bony pelvis.

The lower limb specializes in supporting the body weight and also the locomotion. The muscles, which act upon the lower limb, originate in the pelvic girdle, sacrum, and the vertebral column. It is customary to include the transitional areas when describing the lower limb, that is, the inguinal and gluteal regions.

During development, the lower limbs first appear as minute buds in approximately 5 mm long embryo. Therefore the lower limb buds typically appear approximately 4 weeks postovulation. The lower limb buds are a little bit behind in development as compared to the upper limb buds during development. Each of the lower limb buds lengthens and develops in a proximal to distal sequence, that is, the leg appears before the development of the foot. Just a few days after the development of the lower limb buds, the nerves then start to grow, the skeleton and the muscles become differentiated. Shortly after this, the toes start to form. The list of the bones of the lower limbs is listed in Table 3.1.

The lower limb after development is ideal for our bipedal locomotion. The weight of the body is transferred through the sacroiliac joints via the vertebral column into the pelvic girdle and then to the hip joint and the femur. The arrangement of the femur is oblique to ensure the knee joint is directly inferior to the rest of the body. This ensures that the weight of the body is then transferred downward to the legs and then to the feet. In women, the femurs are at more of an oblique angle due to the wider pelvis.

Essential Clinically Applied Anatomy of the Peripheral Nervous System in the Limbs
http://dx.doi.org/10.1016/B978-0-12-803062-2.00003-6

Table 3.1 The Bones Within Each of the Territories of the Lower Limb

Region of Lower Limb	Bones
Hip and buttock	Ilium, ischium, pubis, acetabulum, sacrum, coccyx
Hip joint	Acetabulum, femur
Thigh	Femur
Knee joint	Distal femur, patella, proximal tibia, and fibula
Leg	Tibia, fibula
Ankle joint	Talus, distal tibia, and fibula
Foot	*Tarsus* (3) (talus, calcaneus, cuboid, navicular), cuneiforms (medial, intermediate, and lateral); *metatarsus* (5) (metatarsals, numbered from the medial side); *phalanges* (14)

TIP!

The fibula does not articulate with the femur. Therefore, it does not transfer weight or bear the weight of the body. However, it does contribute to the ankle joint.

3.2 CUTANEOUS INNERVATION OF THE LOWER LIMB

The lumbosacral plexus innervates the skin of the lower limb. The region of skin, which is supplied by a single spinal nerve, is referred to as a *dermatome*. The supply of the skin by each of these nerves is retained in life, but is slightly distorted in appearance due to the twisting of the limb and lengthening during its development. Although there are *dermatome maps* which outline where each spinal nerve innervates, these can overlap, except within the *axial line*. The dermatomes of the first to fifth lumbar vertebrae (L1–L5) run as a series of bands which extend from the midline of the trunk posteriorly into the lower limbs inferolaterally and anteriorly.

The first and second sacral vertebral level (S1–S2) dermatomes run downward at the posterior aspect of the lower limb, splitting at the ankle with S2 passing medially and S1 passing laterally at the foot.

The rest of this section will discuss each of the cutaneous nerves in detail, providing hints and tips on how to clinically examine them where relevant, and common related pathologies, which are typically found in the clinical setting. The nerves that will be discussed are the subcostal, iliohypogastric, ilioinguinal, genitofemoral, lateral cutaneous nerve of thigh, anterior cutaneous branches, cutaneous nerve of obturator nerve, posterior cutaneous nerve of thigh, saphenous, superficial fibular, deep fibular, sural, medial plantar, lateral plantar, and cluneal nerves.

3.2.1 Subcostal Nerve

The subcostal nerve originates from the ventral ramus of the last (twelfth) thoracic nerve and passes into the abdomen posterior to the arcuate ligament. It then passes inferolaterally, posterior to the kidney, anterior to the quadratus lumborum, and deep to the fascia of the quadratus lumborum. It then goes through the transversus abdominis passing between it and the internal oblique. The subcostal nerve then enters the rectus sheath, going through its front layer. It then becomes superficial halfway between the pubic symphysis and the umbilicus.

The lateral cutaneous branch of the subcostal nerve courses sandwiched between the internal oblique and the external oblique, emerging superficial superior to the iliac crest. It innervates the skin and the subcutaneous tissue of the gluteal region and also the lateral side of the thigh, only as far as the greater trochanter of the femur though.

The subcostal nerve supplies the transversus abdominis, rectus abdominis, and the pyramidalis, along with some fibers to the peritoneum.

3.2.1.1 Clinical Applications
3.2.1.1.1 Autologous Bone Harvesting from Iliac Crest
With the close relations of the subcostal nerve to the anterior superior iliac spine, it can be at risk from damage during autologous bone harvesting from the iliac crest. It has been shown by Chou et al. (2004) that the subcostal nerve may lie as close as 6 cm from the anterior superior iliac spine. Therefore, if it is damaged during harvesting, it could result in postoperative pain at the site of retrieval of bone from the anterior superior iliac spine. This should be kept in mind for biopsies or harvesting of bone from this site.

3.2.1.1.2 Risk from Laparoscopic Surgery
Trocar placement in the superolateral quadrant can damage several nerves for laparoscopic access to the abdomen. It can result in damage to the subcostal nerve or the ventral rami of the lower thoracic nerves and caudal lower intercostal nerves. However, trocar placement in the lower abdominal wall can affect the iliohypogastric and ilioinguinal nerves; their anatomy is discussed in the subsequent sections. Indeed,

with placement of the trocars, it could even result in paresis of the abdominal musculature. This is not common, but should be kept in mind when undertaking laparoscopic examination of the abdomen, and the relevance of the branching pattern and distribution of these nerves (Van Ramshorst et al., 2009).

3.2.1.1.3 Anterior Abdominal Wall Nerve Blocks

One area that has also been receiving interest is the opportunity to undertake nerve blocks of the subcostal nerve and also the ilioinguinal and iliohypogastric nerves. Nerve blocks of these nerves are undertaken to help aid both intraoperative and postoperative analgesia, and also reduce postoperative requirements for opioid analgesia in major abdominal surgery. They have been shown to be effective in earlier mobilization of the patient (Yarwood and Berrill, 2010).

3.2.2 Iliohypogastric Nerve

The thoracoabdominal nerves, as well as the subcostal nerve, detailed in the previous section, supply the abdominal wall. The ilioinguinal and iliohypogastric nerves also do the same.

The thoracoabdominal nerves originate from the seventh to the eleventh intercostal nerves, leaving their respective intercostal spaces, passing anterolateral between the *transversus abdominis* and *internal oblique* abdominal musculature supplying them, as well as the *external oblique*. They enter the sheath of the rectus, and the branches pass anteriorly to supply the rectus and the overlying skin.

The iliohypogastric and the ilioinguinal nerves arise primarily from the first lumbar nerve. Both the iliohypogastric and ilioinguinal nerves are mainly sensory nerves. The thoracoabdominal nerves each supply a band of skin through its anterior and lateral cutaneous branches. As the nerves overlap, sectioning of a single nerve results in diminished sensation in the area that it supplies.

The iliohypogastric nerve can arise from the twelfth thoracic vertebra (i.e., T12). It passes posterior to the lower portion of the kidney and anterior to the quadratus lumborum. It then pierces through the transversus abdominis (posterior part) superior to the iliac crest. It then divides into two branches – anterior cutaneous branch and lateral cutaneous branch.

The anterior branch passes forward between the oblique muscles, piercing the aponeurosis of external oblique and innervates the skin superior to the pubis. The lateral branch passes through the external and internal oblique muscles supplying the skin over the side of the buttocks.

3.2.3 Ilioinguinal Nerve

The ilioinguinal nerve can arise from either the twelfth thoracic vertebra, that is, T12, or the second lumbar vertebra, that is, L2. It runs a similar course to the iliac crest where, after having pierced the transversus abdominis and the internal oblique, it courses anteriorly to accompany the spermatic cord through the inguinal canal. It then emerges from the superficial ring providing innervation to the thigh, and either anterior scrotal or anterior labial branches.

3.2.4 Genitofemoral Nerve

The genitofemoral nerve typically arises from the second lumbar vertebra (L2) or both the first and second (L1–L2) but occasionally from the third lumbar vertebral level (L3). The genitofemoral nerve arises from the anterior aspect of the psoas muscle and descends on it. It then divides into its genital and femoral branches somewhere above the inguinal ligament, though the point of this is variable. The genital branch then goes through the inguinal canal via the deep ring supplying the cremaster, and then passes downward to the scrotum in males, or the labia majora in females. It then terminates in fine filaments, some of which may go on to supply the thigh. The femoral branch, however, enters the femoral sheath, lateral to the femoral artery, passing anteriorly and supplies the skin over the area of the femoral triangle.

3.2.4.1 Clinical Applications

Each spinal nerve has a distribution called a *dermatome*. This is the area that is supplied by a single spinal nerve by its sensory fibers running in its dorsal root. Therefore, applied to the subcostal, ilioinguinal, iliohypogastric, and genitofemoral nerves, these supply certain areas of skin and can be referred to as supplying a dermatome (single area of skin from a single spinal nerve root), or over several dermatomal regions (several areas of skin by several spinal nerves). However, sectioning of a single spinal nerve rarely will result in complete loss of sensation, or anesthesia. This is because other adjacent spinal nerves will also be carrying fibers from that site. Therefore the more likely scenario is a reduced level of sensation, or hypoesthesia.

There are many versions of *dermatome maps* that can be used clinically to test for sensation, to determine where to test for them, and thus to demonstrate where a pathology exists in a patient with a suspected spinal nerve lesion. The most common one to be used clinically is the American Spinal Injury Association's (ASIA) worksheet produced as the International Standards for Neurological Classification of Spinal Cord Injury (ISNCSCI, 2015).

No matter how nerve fibers arrive in an area, the region of skin supplied by a particular segment of the spinal cord is called a dermatome. Dermatomes are rather approximate, and there is often considerable overlap between adjacent dermatomes. Nevertheless, loss of feeling (anesthesia) in a particular dermatome area could indicate that there was damage or disease of the spinal cord segment (or spinal nerve/roots) supplying that area. Increased or abnormal sensation (often painful) could similarly be evoked, for example, by a "slipped disc" pressing on nerves/nerve roots. Altered sensation over the area of distribution of a named peripheral nerve (rather than a dermatome) would suggest damage to a nerve distal to a plexus.

> **TIP!**
>
> A dermatome is a region of skin supplied by one single spinal nerve. It allows you to determine what spinal level pathology may be at. The distribution of our dermatomes can be slightly variable but can give an indication at what level approximately the pathology may exist, for example, upper cervical, lower thoracic, and so on.

Fine touch

> **TIP!**
>
> When examining a patient's sensory system, do not provide suggestions to them as it may influence their interpretation of the examination. DO NOT tell the patient if they notice any changes during the examination, as this suggests that they should expect to notice a change.

3.2.4.2 Light Touch Examination

The purpose of this is to serve as an introductory examination to identify any area or region where pathology may exist in the neurological system. Simple testing should first be done and as the extremities are easy to access without causing too much discomfort to the patient, these should be examined first. The abdomen is easily accessible, whilst maintaining patient decency during the examination. However, the patient may be asked if they want a chaperone present when examining.

> **TIP!**
>
> Testing of *light touch sensation* to identify any areas of abnormality MUST be done when the patient's eyes are closed. This ensures that they do not anticipate where the stimulus is applied. Equally, you MUST tell the patient what you are going to do before doing it. This helps build trust with the patient but also allows them to be fully informed about what you are doing, and that they consent to it.

1. Introduce yourself to the patient stating who you are, and in what capacity and grade you are functioning, for example, student, physician, surgeon, therapist etc.
2. Advise the patient that you want to test for sensation, for example, of the anterior abdominal wall, initially when their eyes are closed.
3. Tell them that you will touch them with either cotton wool, or using a specialist Von Frey filament on various regions of the arms, hands, legs and feet on both the left and right hand sides. You do not need to use the term Von Frey but inform them that it is material made of nylon that allows you to test sensation. It may feel "tickly".
4. Ask the patient if they have understood what you have said and answer any questions they may have.
5. Now, ask the patient to close their eyes and to say "yes" every time they feel you touch, and also to report any feelings of discomfort, pain or abnormal sensations.
6. Working systematically with the cotton wool or Von Frey fibers, touch all dermatome regions of the abdomen, both on the left and right hand sides.
7. Try to compare both the left and right hand sides of each dermatome region. It may not be possible to test for this formally as the patient may find it uncomfortable or rather tickly. Therefore, it may be necessary to localize the testing to a limited number of sites.
8. It will probably not be necessary to include two-point discrimination, though if clinically indicated it can be examined.

If you identify a level, or levels, where there appears to be an alteration in sensation, identify exactly at what point this occurs. When recording the findings from the clinical examination, it can be used on the American Spinal Injury Association's (ASIA) worksheet produced as the International Standards for Neurological Classification of Spinal Cord Injury (ISNCSCI, 2015).

This sheet allows for assessment and classification of motor and sensory function and is classified as follows: *Sensory assessment.*

Light touch and pinprick are assessed separately and a score out of two is given for each dermatome on the right and left side of the body.

The dermatomes that are assessed are cervical (C2, C3, C4, C5, C6, C7, C8), thoracic (T1, T2, T3, T4, T5, T6, T7, T8, T9, T10, T11, T12), lumbar (L1, L2, L3, L4, L5) and sacral (S1, S2, S3, S4, S5).

For sensation the following scoring system is used in Table 3.2.

Table 3.2 The Scoring System for Classification of Sensory Assessments	
Sensory Score	Classification
0	Absent
1	Altered
2	Normal
NT	Not testable

A score is given for light touch right (*LTR* totaling 56, i.e., 2 for each of the vertebral levels stated previously) and light touch left (*LTL* totaling 56, i.e., 2 for each of the vertebral levels stated previously). This is recorded as follows:

$$LTR + LTL = ...\text{out of } 112$$

A score is also given for pinprick right (*PPR* totaling 56, i.e., 2 for each of the vertebral levels stated previously) and pinprick left (*PPL* totaling 56, i.e., 2 for each of the vertebral levels stated previously). This is recorded as follows:

$$PPR + PPL = ...\text{out of } 112$$

3.2.5 Lateral Cutaneous Nerve of Thigh

The lateral cutaneous nerve of thigh, or *lateral femoral cutaneous nerve*, originates from the second and third lumbar vertebrae (L2–L3). On leaving the spinal cord, the nerve then passes from the lateral aspect of the psoas major inferolaterally toward the anterior superior iliac spine, crossing the iliacus. It then passes either deep or superficial to the inguinal ligament, or through it, then to the thigh. However, in the vast majority of cases, the nerve passes deep to the inguinal ligament, before going on to give one to four branches (Zhu et al., 2012). The nerve then enters a compartment between the tensor fasciae latae muscle and the sartorius, which is formed by the fascia lata as a double layer (Patijn et al., 2011).

Hospodar et al. (1999) showed that the course of the lateral cutaneous nerve of thigh had a variable course. They demonstrated that the nerve was typically found 10–15 mm from the anterior superior iliac spine, but could be found as much as 46 mm medially.

3.2.5.1 Clinical Applications

With the variable course of the lateral cutaneous nerve of thigh, its course should be known especially when undertaking any type of needle insertion of the anterior superior iliac spine, for example, harvesting bone grafts from here (Carai et al., 2009).

In addition to this, compression or entrapment of the lateral cutaneous nerve of thigh can occur between the ilium and the attachment point of the inguinal ligament at the anterior superior iliac spine. It results in pain and/or paresthesia or hypoesthesia on the lateral aspect of the thigh, with aching in the groin as well as hypersensitivity to heat.

This condition is called *meralgia paresthetica*, or *Bernhardt-Roth syndrome*. The incidence is approximately 4 per 10,000 person years (van Slobbe et al., 2004). Entrapment of the nerve is caused by tight clothing (Moucharafieh et al., 2008), or by sports like cycling, or long distance walking. In addition, it can be caused by pregnancy or change in weight (Pearce, 2006). Treatment of this condition can either be conservative, or involve neurectomy for severe cases, but that will result in numbness in the distribution of the nerve. It may actually be better to have a lack of sensation in that area than the pain experienced in some patients.

3.2.5.1.1 Anterior Cutaneous Branches

The femoral nerve arises from the second to fourth lumbar vertebral levels and is the largest of the branches of the lumbar plexus. It arises within the substance of the psoas major muscle and emerges from its lateral border just inferior to the iliac crest. It passes between iliacus and psoas major passing into the thigh deep to the inguinal ligament.

As mentioned earlier, it gives off the lateral (femoral) cutaneous nerve. It then supplies the iliacus, pectineus, sartorius, and the quadriceps muscles (vastus lateralis, intermedius and medialis, and rectus femoris). It also provides articular branches to both the hip and knee joint.

The anterior cutaneous branches (of the femoral nerve) arise from the second to fourth lumbar vertebrae from the femoral nerve (L2–L4). The anterior cutaneous branches arise within the femoral triangle and pass through the fascia lata alongside the sartorius. The anterior cutaneous branches of the femoral nerve are the intermediate and medial cutaneous nerves. The intermediate femoral cutaneous nerve is typically two branches. The branches give muscular twigs to the sartorius and then pierce the fascia to descend on the front of the thigh. This supplies the skin and contributes to the patellar plexus. The medial cutaneous nerve crosses superficial to the femoral vessels (artery and vein) at the apex of the femoral triangle. These branches supply the skin on the medial side of the thigh contributing to the subsartorial and patellar plexuses. The branches therefore of the anterior cutaneous nerve (from the femoral nerve) are distributed to the skin of the medial and anterior regions of the thigh.

3.2.6 Cutaneous Nerve of Obturator Nerve

The obturator nerve originates from the third and fourth lumbar vertebrae (L3–L4), emerging from psoas major at its medial border. It accompanies the obturator vessels to the obturator groove, where it then divides into anterior and posterior branches.

The anterior branch supplies the gracilis, adductor longus and brevis, and sometimes pectineus. The posterior branch pierces the obturator externus descending anterior to the adductor magnus and posterior to the adductor brevis, supplying these muscles as well as the obturator externus.

The cutaneous branch of the obturator nerve arises from the anterior division of the obturator nerve and has a variable distribution to the skin of the medial aspect of the thigh. Through operative and MRI scans, Bouaziz et al. (2002) had shown that almost half of the patients that they studied had no cutaneous innervation from the obturator nerve. This therefore has relevance in three-in-one nerve blocks of the femoral, lateral femoral cutaneous, and obturator nerves when trying to ensure anesthesia of these nerves for knee surgery (Lang et al., 1993; Bouaziz et al., 2002).

3.2.6.1 Clinical Examination
Testing of the cutaneous branch of the obturator nerve can be undertaken by assessing sensation at the medial aspect of the thigh around the knee joint. Typical things to perform when testing for sensation of nerves is given in more detail in the earlier Section 3.2.4.

During the three-in-one nerve block previously described, another thing to test for clinically is assessment of adductor muscle strength. This is because the three-in-one nerve block will paralyze, or weaken, the adductor muscles, which are also supplied by the main obturator nerve. The strength of the muscles can be tested using a sphygmomanometer slightly inflated (e.g., 40 mmHg); with the patient's hips and knees flexed, squeeze the sphygmomanometer to give an indication of adductor muscle strength (Bouaziz et al., 2002).

3.2.6.2 Clinical Applications
With patients that have a femoral fracture, or need anesthesia to reduce postoperative pain after hip repair and replacement, a three-in-one nerve block can be performed under ultrasound guidance. This can therefore provide appropriate pain relief and reduce the amount of opioids given to these patients (Christos et al., 2010). The nerves which are anesthetized are the femoral, obturator, and lateral cutaneous nerves. Under ultrasound guidance and anesthesia of the injection site, the area around the femoral nerve is anesthetized by injecting "20 mL of 0.5% bupivacaine, which is inserted 2 cm distal to the inguinal ligament in a lateral to medial direction at a 30° angle" (Christos et al., 2010).

3.2.7 Posterior Cutaneous Nerve of Thigh
The posterior cutaneous nerve of thigh, also called the *posterior femoral cutaneous nerve*, is from the first three sacral vertebrae (S1–S3). It enters the gluteal region through the greater sciatic foramen, inferior to the piriformis. It descends deep to the gluteus maximus along with

the sciatic nerve and the inferior gluteal artery. Close to the popliteal fossa it pierces the fascia and then runs with the small saphenous vein to the mid-calf level and its smaller fibers then merge with the sural nerve.

When it is deep to the gluteus maximus, the posterior cutaneous nerve of thigh gives off *inferior clunial nerves* (gluteal branches). These branches wind round the inferior border of the gluteus maximus and then innervate the skin of the buttocks.

At the inferior border of the gluteus maximus, the posterior cutaneous nerve of thigh gives *perineal branches*. These cross the hamstring muscles and innervate the skin of the external genitalia.

Femoral and *sural* branches are also given off at intervals innervating the skin on the posterior aspect of the thigh and calf, with some reaching the heel.

3.2.7.1 Clinical Examination
Testing of the posterior cutaneous nerve of thigh can be undertaken by assessing sensation over the buttocks and the posterior aspect of the thigh and calf. Typical things to perform when testing for sensation of nerves are given in more detail in the earlier Section 3.2.4, on the genitofemoral nerve.

3.2.7.2 Clinical Applications
For those patients that may require the application of a tourniquet to the lower limb, especially in cases of trauma, a posterior femoral cutaneous nerve block may be required. This would provide anesthesia to the knee, leg and foot (Topçu and Aysel, 2014).

3.2.8 Saphenous Nerve
The saphenous nerve originates from the third and fourth lumbar vertebral levels (L3–L4) and is the termination of the femoral nerve. It descends with the femoral artery and vein through the femoral triangle and then onto the subsartorial canal. In the canal it crosses the femoral artery passing medially. Then with the saphenous branch of the descending genicular artery, it becomes superficial between the gracilis and the sartorius. The saphenous nerve descends in the leg along with the large (or great) saphenous vein supplying the skin of the medial side of

the leg and foot. It also gives branches to the knee joint and contributes to the patellar and subsartorial plexus.

It has also been shown that the saphenous nerve has two main branches – anterior and posterior (Mercer et al., 2011). These branches arise roughly 3 cm above the medial malleolus. The anterior branch of the saphenous nerve terminates at the front of the medial malleolus close to the posterior aspect of the great (long) saphenous vein. The posterior branch of the saphenous nerve terminates at the posterior edge of the medial malleolus.

3.2.8.1 Clinical Applications
3.2.8.1.1 Saphenous Vein Cutdown

The insertion of a catheter into peripheral veins is a common clinical procedure used to obtain blood, and provide a temporary means for giving drugs into the vascular system for a quicker acting response. With patients who have suffered trauma and also have hypovolemic shock, peripheral veins typically used in the cubital fossa may not be as easy to access. Therefore, a *saphenous vein cutdown* procedure can be undertaken.

The *great saphenous vein*, also called the *large*, long, or greater saphenous vein, starts at the junction of the dorsal digital vein of the medial side of the great toe with the dorsal venous arch. It then ascends the leg by passing in front of the medial malleolus crossing the medial side of the tibia obliquely, along with the saphenous nerve. It then ascends the medial border of the tibia, passing posterior to the medial condyles of the tibia and the femur. The greater saphenous vein then continues its ascent along the medial side of the thigh toward the femoral triangle. It then pierces the cribriform fascia, which occupies the saphenous opening in the fascia lata. Further, it pierces the femoral sheath terminating in the femoral vein.

For a saphenous vein cutdown, it is accessed at the ankle with a horizontal incision approximately 1–2 cm in front to the medial malleolus. The distal aspect of the vein is ligated and a catheter is inserted, followed by it being secured in place.

With the close relationship of the saphenous nerve and this vein, it may place the saphenous nerve at risk of being sectioned. This would result in the patient *losing cutaneous innervation on the medial aspect of the leg* and should be kept in mind when undertaking this procedure.

3.2.8.1.2 Harvesting of the Great Saphenous Vein

With patients who have coronary artery disease (CAD), and who require coronary artery bypass surgery (CABG), it may be necessary to take the great saphenous vein to use as a bypass graft material, though the internal thoracic (mammary) artery can also be used.

One of the issues when harvesting the great saphenous vein is injury to the saphenous nerve due to its intimate relationship with this structure. However, rather than the traditional open technique, which involved making an incision from the medial malleolus to the groin, a preference was for a standard bridging technique (SBT). The SBT involved making smaller incisions along the length of the course of the great saphenous vein; doing it this way also results in saphenous nerve injury and neuropathy (Khan et al., 2010).

Endoscopic access to the great saphenous vein has also received some attention, but can also be associated with lower patency rates (Zenati et al., 2010).

3.2.8.1.3 Saphenous Nerve Block

Saphenous nerve block can be undertaken for chronic knee pain associated with, for example, osteoarthritis (Lotero et al., 2014), or for reduction of postoperative pain in total knee arthroplasty (Andersen et al., 2013). This can be undertaken with ultrasound guidance to ensure accuracy of the infiltration of the anesthetic (Lotero et al., 2014)

3.2.9 Superficial Fibular Nerve

The superficial fibular nerve, also called the superficial peroneal nerve, is one of the terminal branches of the common peroneal nerve. It descends anterior to the fibula and between the perinei (fibularis muscles) and the extensor digitorum longus. In the lower leg, the superficial fibular (peroneal) nerve divides into medial and intermediate dorsal cutaneous branches, which pass anterior to the extensor retinacula to supply the toes.

Muscular branches of this nerve pass to the peroneus (fibularis) longus and brevis. The branch of the nerve going to peroneus (fibularis) brevis often extends to the lateral malleolus and sends small branches to the extensor digitorum brevis and nearby joints and ligaments.

The two terminal cutaneous branches supply the anterior lower one-third of the leg and the dorsum of the foot. On the medial side, the *medial dorsal cutaneous nerve* supplies the medial side of the big toes, gives a branch to the deep peroneal nerve, and provides a branch which divides into dorsal digital branches for the second and third toes lying adjacent to it. The more lateral of the branches is called the *intermediate dorsal cutaneous nerve*. It divides into two branches, which divide into dorsal digital nerves for the third through to the fifth toes. The cutaneous nerves are highly variable in their distribution. The plantar digital branches from the medial and lateral plantar nerves, which are discussed in Sections 3.2.12 and 3.2.13 respectively, innervate the nails and the tips of the toes.

3.2.9.1 Clinical Examination
Testing of the superficial fibular (peroneal) nerve can be undertaken by assessing sensation over the anterolateral region of the leg as well as the medial half of the big (great) toe, lateral aspect of the second toes, third and fourth toes and the medial side of the fifth toe. Note that the deep fibular (peroneal) nerve innervates the web space between the big and second toes (i.e., lateral aspect of the big toe and medial side of the second toe). The sural nerve innervates the lateral aspect of the fifth toe. Typical things to perform when testing for sensation of nerves are given in more detail in the earlier Section 3.2.4, on the genitofemoral nerve.

3.2.9.2 Clinical Applications
3.2.9.2.1 Superficial Fibular (Peroneal) Nerve Entrapment
Entrapment of nerves tends to occur at points where nerves change direction or pass over bony prominences. In addition, a variety of other causes can cause nerve entrapment like plaster casts, operative procedures at or near the site of the nerve, ganglions, cysts, or lipomas (Hammer, 2007). As the superficial fibular nerve arises from the common fibular/peroneal nerve and crosses the head of the fibula, this site is prone to damage during trauma or simply sitting with the legs crossed for long periods of time. It results in a peroneal neuropathy, and can affect the superficial fibular (peroneal) nerve too.

Symptoms of superficial fibular (peroneal) nerve entrapment include pain, numbness and paresthesia in the distribution of the nerve, that is, down the lateral aspect of the leg (calf) and also the dorsum of the foot, apart from the web space between the first and second toes.

In addition, superficial fibular (peroneal) nerve entrapment can also result in compression within the crural fascia, and therefore treatment of this relatively uncommon neuropathy has to be directed at the cause (Paraskevas et al., 2014), with neurolysis being the best treatment in more extreme cases after nerve repair has been undertaken (Baima and Krivickas, 2008).

3.2.10 Deep Fibular Nerve

The deep fibular nerve is also referred to as the deep peroneal nerve. As with the superficial fibular/peroneal nerve, it originates at the bifurcation of the common peroneal nerve and is its terminal branch. It continues winding around the neck of the fibula under the peroneus longus. It then pierces the anterior intermuscular septum and extensor digitorum longus, passing inferiorly on the intermuscular membrane, under the cover of extensor hallucis longus and extensor digitorum longus. It then appears close to the anterior tibial artery initially found lateral to this vessel, then anterior, and then it is found lateral to this vessel again. Both the anterior tibial artery and the deep peroneal nerve are found deep to the extensor retinaculum. The final branches of the deep peroneal nerve are the lateral and medial branches, like the superficial peroneal nerve.

The lateral branch courses over the tarsus, deep to the extensor digitorum brevis, supplying it as it does so. In addition, the lateral branch supplies the 2nd–4th toes for their tarsal and metatarsophalangeal joints. It also supplies the second dorsal interosseous muscle.

The medial branch of the deep peroneal nerve supplies the lateral aspect of the great toe and the medial aspect of the second toe. It also supplies the metatarsophalangeal joint of the great toe, and also gives off a twig to the first dorsal interosseous muscle. Damage to this branch of the common peroneal (fibular) nerve can result in footdrop due to its innervation of muscles in the anterior compartment of the leg (see Section 3.3.13 for further information). Indeed, there would also be loss or reduced sensation in the first web space.

3.2.10.1 Clinical Applications

3.2.10.1.1 Deep Fibular Nerve Entrapment

As with the superficial fibular (peroneal) nerve, the deep fibular (peroneal) nerve is also subject to nerve entrapment pathology, and perhaps under diagnosed.

The nerve supplies the extensor muscles of the leg and the sensory branch provides innervation to the web space between the big (first) and second toes. It innervates the tibialis anterior, extensor digitorum longus, extensor hallucis longus, and peroneus tertius muscles. At approximately 1.5 cm superior to the ankle joint, it provides innervation to the extensor digitorum brevis and sensory supply to the ankle joint.

Modern means of imaging means that damage to this branch is best seen using MRI. Typically, it is much easier to identify in the older patient, and those with more adipose tissue compared to more athletic patients. This is because the muscle bulk in more athletic patients may mask the "fat stripe" in front of the interosseous membrane. It has also been demonstrated that the deep fibular (peroneal) nerve could be seen easily at the ankle within the dorsal fat, and deep to the extensor retinaculum, between the tendons of extensor digitorum longus and the extensor hallucis longus. They also showed that short-axis imaging in the mid-foot region are best for tracing the medial and lateral final branches of the deep fibular (peroneal) nerve.

Typically the deep fibular (peroneal) nerve is prone to injury at the tarsal tunnel and also at the dorsum of the foot (Marinacci, 1968). It can be damaged in athletes including skiers (Lindenbaum, 1979), those who play soccer (Borges et al., 1981), and also in figure skaters (Melendez et al., 2013).

Again, treatment of entrapment of the deep fibular (peroneal) nerve depends on the cause, and may require surgical intervention to release the nerve from pressure from surrounding structures.

3.2.11 Sural Nerve

The sural nerve is comprised of four different parts – the lateral and medial sural cutaneous nerves, the peroneal communicating branch and the sural nerve itself.

- *Lateral sural cutaneous nerve*: The lateral sural cutaneous nerve typically arises from the common peroneal nerve, and then gives off the peroneal communicating branch approximately 3–8.5 cm after leaving the common peroneal nerve (Ortigüela et al., 1987). It typically splits to form a lateral and medial division with the medial one passing toward the midline posteriorly. The lateral branch terminates in the skin of the lateral aspect of the leg.
- *Medial sural cutaneous nerve*: The medial sural cutaneous nerve arises from the tibial nerve at its posterior aspect. It is found in the popliteal fossa close to the origin of the tibial nerve. The medial sural cutaneous nerve then passes deep to the fascia of the gastrocnemius, between its two heads. This nerve then emerges superficially by passing through the crural fascia roughly at the middle to distal one third of the leg, then joining with the sural nerve (Ortigüela et al., 1987).
- *Peroneal communicating branch*: The peroneal communicating branch to the sural nerve arises from the lateral sural cutaneous nerve, as previously mentioned.

The sural nerve itself is formed from the medial and lateral sural cutaneous nerves, from the tibial and the common peroneal (fibular) nerves respectively. However, the sural nerve has also been found to arise from the medial sural cutaneous nerve and the peroneal communicating branch, or simply a continuation from the medial sural cutaneous nerve.

The sural nerve passes distally close to the short (or lesser) saphenous vein, passing approximately 1–1.5 cm behind the lateral malleolus. It innervates the lateral aspect of the little (fifth) toe. Ortigüela et al. (1987) also demonstrated that there are sometimes two or three small branches to the skin just proximal to the ankle.

3.2.11.1 Clinical Applications

Due to the superficial location of the sural nerve, it lends itself to be an ideal candidate for nerve biopsy in diagnosing neural disease, and can also be used for nerve grafting procedures where there are defects in tissue resulting from wounds (Moore and Dalley, 2006). It has been shown that harvesting of this graft only leaves mild residual symptoms similar to biopsy at this site (Hallgren et al., 2013).

Table 3.3 The Structures Supplied by the Medial Plantar Nerve	
Medial plantar nerve	
Muscular branches	Flexor hallucis brevis, flexor digitorum brevis, abductor hallucis, 1st lumbrical
Cutaneous innervation	Medial three and one-half toes, nail beds and tips (dorsal)

3.2.12 Medial Plantar Nerve

The nerves of the foot include the superficial and deep fibular (peroneal) nerves as well as the lateral and medial plantar nerves.

The medial plantar nerve is the larger of the terminal branches of the tibial nerve, and arises from under the cover of the flexor retinaculum. Initially the medial plantar nerve is deep to the adductor hallucis, and then it passes anterior in the sole and then is found sandwiched between the flexor digitorum brevis and the abductor hallucis. The medial plantar nerve lies lateral to the medial plantar artery. Table 3.3 highlights the muscular and cutaneous distribution of the medial plantar nerve.

3.2.13 Lateral Plantar Nerve

The lateral plantar nerve is the other terminal branch of the tibial nerve. It arises from below the flexor retinaculum and passes anterior, deep to the abductor hallucis and flexor digitorum brevis. On its course, the lateral plantar nerve is accompanied by the lateral plantar artery (from the posterior tibial artery), which is on the lateral aspect it.

When it arrives at the base of the fifth metatarsal, it then subdivides into a deep and superficial branch. The individual branches are listed in Table 3.4 demonstrating the main trunk and the superficial and deep branches, as well as the motor supply and cutaneous distribution, as relevant.

3.2.14 Clunial Nerves

The clunial (cluneal) nerves are nerves that innervate the skin overlying the buttocks. As such, there are three major divisions – the superior, medial, and inferior clunial (cluneal) nerves.

1. *Superior clunial (cluneal) nerve*: The superior clunial (cluneal) nerves arise from the posterior rami of the first to third lumbar vertebral

Table 3.4 The Branches of the Lateral Plantar Nerve, and the Related Subdivisions and what they Supply in Terms of Motor and Sensory Innervation

Branch Origin	Innervation
Main trunk	*Motor*: quadratus plantae, abductor digiti minimi; *sensory*: cutaneous innervation of the lateral part of the sole
Superficial branch	*Motor*: flexor digiti minimi brevis, interosseous muscles of fourth intermetatarsal space; *cutaneous*: sides of the lateral one and one-half toes, nail beds and tips of toes (dorsal)
Deep branch	*Motor*: adductor hallucis, 2nd–4th lumbricals, all interossei (except those in the fourth intermetatarsal space), articular twigs

levels (L1–L3). After originating from L1 to L3, the superior clunial (cluneal) passes in close relationship to the iliac crest and then runs through a tunnel formed by the thoracodorsal fascia and the iliac crest. It then runs inferolaterally within the subcutaneous tissue. The superior clunial (cluneal) nerve innervates the skin on the central and superior aspects of the buttocks.

2. *Medial clunial (cluneal) nerve*: The medial clunial (cluneal) nerve originates from the first to the third sacral vertebral levels (S1–S3). It arises from the dorsal sacral foramina and then innervates the skin of the medial side of the buttocks and also the intergluteal cleft, at the midline.

3. *Inferior clunial (cluneal) nerve*: The inferior clunial (cluneal) nerve originates from the *posterior cutaneous nerve of the thigh*. The posterior cutaneous nerve of the thigh, also called the *posterior femoral cutaneous nerve*, is from the first three sacral vertebrae (S1–S3). It enters the gluteal region through the greater sciatic foramen, inferior to the piriformis. It descends deep to the gluteus maximus along with the sciatic nerve and the inferior gluteal artery. Close to the popliteal fossa it pierces the fascia and then runs with the small saphenous vein to the mid-calf level and its smaller fibers then merge with the sural nerve.

When it is deep to the gluteus maximus, the posterior cutaneous nerve of thigh gives off *inferior clunial (cluneal) nerves* (gluteal branches). These branches wind round the inferior border of the gluteus maximus and then onward to innervate the skin of the buttocks. Specifically, the inferior clunial (cluneal) nerve innervates the skin of the inferior buttock, generally over the gluteal fold.

3.2.14.1 Clinical Applications

3.2.14.1.1 Injury to the Clunial (Cluneal) Nerves

Injury to the clunial (cluneal) nerves can occur during harvesting of autologous bone grafting from the iliac crest. Indeed, obtaining bone from the posterior iliac crest has resulted in postoperative pain from that site being the most common complication (Delawi et al., 2007).

In addition to this, the clunial nerves can be susceptible to being damaged during surgical procedures carried out on the lumbar discs where there was fusion using the iliac crest (Frymoyer et al., 1978).

With bone grafting from the iliac crest including for lumbar fusion, it could result in damage to the superior clunial nerves. This would result in pain at the site of the graft, and most likely would be due to damage of the superior clunial nerves (Berthelot et al., 1996). As with injury to other peripheral cutaneous nerves, this can present in a variety of ways including abnormal sensations like hypoesthesia, hyperesthesia, or indeed long term pain.

A variety of approaches to harvesting from, and access to the posterior iliac spine (Colterjohn and Bednar, 1997) had shown that if incisions were made parallel to the pathway of the superior clunial (cluneal) nerves, and at right angles to the posterior aspect of the iliac crest, this would reduce the incidence of injury to these nerves.

Using a straight line 2.5 cm in front of the posterior superior iliac spine and at right angles to the long axis of the posterior iliac spine has been recommended. They also advised that a subperiosteal dissection over the region of the posterior superior iliac spine at a distance of 2.5 cm on either side of the axis would help avoid not only the superior clunial (cluneal) nerve but also the middle/medial clunial (cluneal) nerve.

In relation to measurements, Colterjohn and Bednar (1997) advised that the superior clunial (cluneal) nerve crossed the iliac crest 8 cm away from the midline and 7 cm cephalad to the posterior superior iliac spine.

3.2.14.1.2 Clunial Nerve Entrapment Neuropathy

Strong and Davila (1957) were the first to describe a *clunial nerve syndrome* that presented as pain in the lower back region; in many cases, the pain radiated down the lower limb.

Maigne (1980) and Maigne et al. (1989) followed this up and defined it as *Maigne's syndrome*. Typically this would involve facet syndromes in the region of the junction between the thorax and lumbar vertebrae resulting in lower back pain. In Maigne's syndrome the pain is not felt at this junction but more inferiorly. Typically, the pain would be referred further down. Pressure on the iliac crest would bring on this pain at the lower back, corresponding to the sensory distribution of the superior clunial (cluneal) nerve.

Maigne et al. (1989) also stated that there could be entrapment of the superior clunial (cluneal) nerve within an osteofibrous tunnel within the space enclosed by the iliac crest and that of the thoracolumbar fascia for those patients presenting with lower back pain.

Recently Kuniya et al. (2013) revealed through an extensive clinical study that entrapment of the superior clunial nerve was perhaps not as uncommon as initially suspected, for example, by Talu et al. (2000). They studied a total of 834 patients who had lower back pain with or without lower limb related pain. They used the diagnostic criteria of involvement of the superior clunial (cluneal) nerve as pain, which was most severe on the iliac crest posteriorly, at 7 cm from the midline, with point tenderness at that specific point. They performed nerve blocks on those patients and used a visual analog score to assess the effectiveness of the nerve blocks, although they omitted the details of these scales and the outcomes from their research paper.

They demonstrated that, of those patients that met their criteria, the visual analog scale was significantly reduced by nerve blocks targeted at the superior clunial (cluneal) nerves. They stated that pathology or nerve entrapment of the superior clunial (cluneal) nerves should always be suspected in patient's with long standing lower back pain, perhaps radiating to the lower limb, but without any obvious organic or structural cause.

3.3 MOTOR AND SENSORY NERVES OF THE LOWER LIMB

3.3.1 Superior Gluteal Nerve

The superior gluteal nerve originates from the sacral plexus, specifically from the dorsal branches of the ventral rami of lumbar nerves four and five, as well as the first sacral nerve. The origin of the superior gluteal nerve was found to be dorsal and more proximal to the origin of the inferior gluteal nerve (Akita et al., 1992). The superior gluteal nerve then runs posterolaterally leaving the pelvis via the greater sciatic foramen. It is closely related to the superior aspect of piriformis, and at this point, the superior gluteal vessels also accompany it. The superior gluteal artery is the biggest of the branches of the internal iliac artery (from the common iliac artery). The superior gluteal veins are the accompanying veins of the superior gluteal artery and drain to the hypogastric, or internal iliac vein.

Superior to the piriformis, the superior gluteal nerve then divides into two divisions – a superior branch and an inferior branch. The superior branch of the superior gluteal nerve runs with the superior gluteal artery's upper portion of the deep division. The superior branch of the superior gluteal nerve innervates the gluteus medius and sometimes provides innervation to the gluteus minimus. The superior gluteal nerve passes across the deep aspect of the gluteus medius at approximately 5 cm superior to the tip of the greater trochanter (Jacobs and Buxton, 1989).

The inferior branch of the superior gluteal nerve accompanies the lower part of the deep division of the superior gluteal artery. The inferior branch of the nerve crosses the gluteus minimus providing innervation of this muscle, but also innervates the gluteus medius. The inferior branch of the superior gluteal nerve then terminates by innervating the tensor fasciae latae.

1. *Gluteus Medius*: The gluteus medius (Figure 3.1) is a rather thick muscle which is fan shaped. It originates from the outer aspect of the ilium from the iliac crest superiorly, the middle gluteal nerve inferiorly and the posterior gluteal line. It then runs inferolaterally to pass to the lateral surface of the greater trochanter.
   The gluteus medius is responsible for strong abduction at the hip joint working in tandem with the gluteus minimus and also the tensor fasciae latae. The more anterior fibers of the gluteus medius are also responsible for medial rotation of the thigh.
   From a functional perspective, the role of the gluteus medius (along with the gluteus minimus and the tensor fasciae latae) comes into

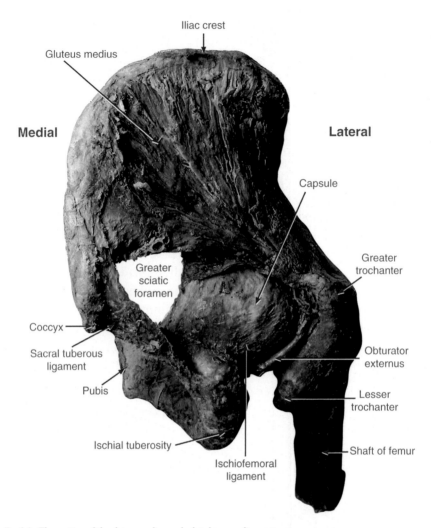

Fig. 3.1. The position of the gluteus medius, and related surrounding anatomy.

action during walking and running. During walking and running, the limb, which is not in contact with the ground during the gait cycle, is prevented from titling downward by the contraction of the gluteus medius (and the gluteus minimus and the tensor fasciae latae) on the opposite side, that is, that limb which is in contact with the ground.

2. *Gluteus Minimus*: Like the gluteus medius, this muscle is also fan-shaped in appearance. It is found deep to the gluteus medius. It arises from the outer area of the ilium between the inferior and middle gluteal lines. The fibers of the gluteus minimus then pass,

like the gluteus medius, inferolaterally to also attach onto the greater trochanter, but more at its anterior surface.

The function of the gluteus minimus is exactly the same as the gluteus medius, that is, abduction of the hip joint, and also medial rotation of the thigh with its more anterior portion.

3. *Tensor Fasciae Latae*: The tensor fasciae latae is a small muscle, which arises from the outer aspect of the iliac crest, located between the iliac tubercle and the anterior superior iliac spine. It then passes inferiorly and posteriorly and is within a sheath formed from the iliotibial tract of the fascia lata (on the lateral aspect of the thigh). It forms the superior aspect of the iliotibial tract.

The tensor fasciae latae provides traction on the iliotibial tract. Therefore, it helps with the action of the gluteus maximus in ensuring the knee is in the extended position. In other words, provided that the iliotibial tract stays anterior to the axis of flexion of the knee, it will ensure the knee is straight. When standing, this force superiorly from the iliotibial tract is the key component to maintain the knee in the fully extended position.

3.3.1.1 Clinical Applications

When undertaking any clinical history taking or examination, you should always do the following:

1. Introduce yourself to the patient.
2. Advise them of what position you hold, for example, student, specialty grade, consultant etc.
3. Your reason for consulting with them, or to find out why they have presented to you.
4. Always take a thorough and detailed history, which will be guided by the presenting signs and symptoms.
5. When examining the patient, always tell them what you will ask them to do, or what region of the body you will be examining, with specific instructions and ensure they give consent.

3.3.1.1.1 Testing of the Superior Gluteal Nerve
- If the patient is able to stand unaided, ask the patient to stand with their feet together.
- Then, ask the patient to raise one foot from the ground, while balancing on the other. Always ensure that the patient is relatively steady, and does not show obvious signs of falling.

- Repeat this on the opposite side.
- The hips should be in a relatively constant position on both sides when lifting one foot from the ground, both on the raised side, and the side on the ground. There may be a slight raising of the unsupported hip which is normal.

3.3.1.1.2 Superior Gluteal Nerve Injury

Injury to the superior gluteal nerve can happen due to *dislocation of the hip joint, hip fractures, repair of hip fractures, and also intramuscular injection in the buttocks* (Bos et al., 1994; Comstock et al., 1994; Gulec and Buyukbebeci, 1996; Lavigne and Loriot de Rouvray, 1994; Ogbemudia et al., 2010).

Injury of the superior gluteal nerve will paralyze the innervation to the gluteus medius muscle resulting in a characteristic appearance on walking and standing. On asking the patient to raise their leg from the ground on the suspected side of the pathology, the pelvis on the unsupported side will tip inferiorly. This is due to the lack of contraction of the gluteus medius (and minimus), which helps stabilize the pelvis. This appearance is referred to as the *Trendelenburg sign*. It was first described by German Friedrich Trendelenburg (1895), who described the sign in relation to patients with congenital dislocation of the hip and also progressive muscular atrophy.

In addition to this, the patient will also have a characteristic "waddling" or gluteal gait. As the pelvis descends opposite to the side of the pathology, the lower limb, in essence, becomes too long when approaching the ground in walking. As a result, the patient will tilt their pelvis to the affected side to raise the foot from the ground. The limb on the unaffected side has to be raised enough to clear the ground during the swing phase of walking. Also, the patient may raise the foot on the unsupported side during walking as it swings forward. This is called the *steppage gait*. If the foot on the unsupported side also swings out laterally, it is called the *swing-out gait*.

3.3.1.1.3 Gluteal Injections

The gluteal region is the site commonly used for the administration of drugs. The reasons for this include:

1. Drug altered or not viable through the intestinal route (e.g., oral or per rectum)
2. Rapid action of the drug required

3. Drug not available in oral form
4. Patient unable to take drugs orally

Each buttock is typically divided into quadrants, and the administration of the injection should be in the superolateral quadrant. This is to avoid not only the branching of the superior gluteal nerve, but also the sciatic nerve, which is found in the lower quadrants.

Additional advice also given to nurses who typically give these injections frequently, if relevant to their clinical practice states, that the superolateral quadrant should be further divided into quadrants again, and the upper outermost one is the site of the injection for the patient (Central Manchester University Hospitals, 2015).

Possible complications of giving an intramuscular injection into the gluteal muscle include injury to the superior gluteal nerve terminal branches, hematoma, abscess formation, pain, infection, lipodystrophy, or intramuscular bleeding (Central Manchester University Hospitals, 2015).

3.3.2 Inferior Gluteal Nerve

The inferior gluteal nerve originates from the dorsal divisions of the ventral rami of the L5–S2 nerves. After leaving the lumbosacral plexus, it exits the pelvis through the greater sciatic foramen and is inferior to the piriformis and medial to the sciatic nerve. It then penetrates the deep surface of the *gluteus maximus* to supply it. There are no sensory fibers in the inferior gluteal nerve. The inferior gluteal nerve accompanies the inferior gluteal artery, and is related to exactly the same structures as the vessel. In other words the inferior gluteal nerve is closely associated with the posterior femoral cutaneous nerve, and lies posterior to the quadratus femoris, adductor magnus, gemelli and the obturator internus. Occasionally, the inferior gluteal nerve may pass through the substance of the piriformis; this rarely occurs bilaterally (Tillmann, 1979).

From a clinically applied anatomical and surgical relevance, especially in posterior approaches to hip surgery, the inferior gluteal nerve has been found to enter the deep surface of the gluteus maximus approximately 5 cm from the tip of the greater trochanter of the femur (Ling and Kumar, 2006).

Gluteus maximus is a very powerful muscle in man. It is a large, thick quadrilateral muscle mass with very coarse fascicles separated by fibrous septae. Of all the muscles in the body, it is the largest and also the most superficial; combined with its adipose tissue superficial to its main bulk, it forms the buttocks. The muscle is an incredibly powerful structure and is used as an antigravity muscle. In other words, it allows us to raise ourselves from the sitting to the standing position and also allows us to climb, for example, stairs, ladders, mountains etc.

The gluteus maximus has extensive attachments. It extends superiorly from the ilium at the posterior gluteal line, iliac crest, dorsal inferior aspect of the sacrum, and coccyx and also the erector spinae aponeurosis. It descends inferolaterally passing over the greater trochanter ending in the iliotibial tract from the tensor fasciae latae. The deep portion of the gluteus maximus attaches onto the *gluteal tuberosity*, which is the *lateral lip of the linea aspera* on the posterior surface of the femur.

The gluteus maximus is responsible for *extension of the thigh* when it acts from the pelvis. It is also active in *external rotation of the hip* as well as *abduction of the thigh* and *lateral rotation of the thigh*. Through its connection with the iliotibial tract, it also supports the knee thus steadying the femur relative to the tibia.

Three bursae (Gk. Byrsa = wineskin) are also associated with this muscle – the *trochanteric*, *gluteofemoral*, and *ischial bursae*. A bursa is a closed synovial fluid-filled sac, which allows reduction of friction (e.g., where a tendon passes over bone). The trochanteric bursa is found over the greater trochanter, separating the muscle from this bony structure. The second, the gluteofemoral bursa, is found between the tendon of gluteus maximus and the vastus lateralis. The third is found over the ischial tuberosity separating the gluteus maximus from the bony landmark.

3.3.2.1 Clinical Examination
One way to assess the function of the gluteus maximus, and therefore its innervation from the inferior gluteal nerve, is the "gluteal skyline test" (White et al., 2009).

Here the patient lies flat and the gluteal region is examined. In pathology of the inferior gluteal nerve, there will be wasting of the gluteus maximus muscle compared to the unaffected side. The muscle bulk will appear wasted.

Then, ask the patient to contract their buttocks. The affected side will be less active and the contraction will be less pronounced or absent. A positive test using this method can suggest a lesion involving the nerve roots of the inferior gluteal nerve, that is, L5–S2.

In addition to this, you can ask the patient to stand from the sitting position, or to walk up a few stairs. Observation of the gluteal region will show sagging of the muscle belly, absent infragluteal fold and the hollow by the greater trochanter disappears. When observing the gluteus maximus territory be sure to compare both sides.

3.3.2.2 Clinical Applications

Ling and Kumar (2006) investigated the anatomy and its relations of the inferior gluteal nerve. They had suggested that the posterior approach for hip replacements could place the nerve at risk from this procedure. Indeed, Abitbol et al. (1990) had shown that 35 out of 45 patients operated using a lateral approach for the total hip replacement had abnormal findings from electromyography studies of superior and inferior gluteal nerve innervations (i.e., gluteus medius and minimus and gluteus maximus respectively).

A variety of approaches have been described when doing the posterior approach to the hip joint. However, Ling and Kumar (2006) suggest that the inferior gluteal nerve enters on the deep surface of the gluteus maximus roughly 5 cm away from the tip of the greater trochanter. They state that the inferior gluteal nerve may therefore be damaged if the gluteus maximus muscle is split during access to the posterior aspect of the hip joint, more than 5 cm from the tip of the greater trochanter. Ling and Kumar (2006) therefore suggest splitting the gluteus maximus muscle not more than 5 cm from the tip of the greater trochanter on the femur toward the posterior superior iliac spine.

> **TIP!**
>
> The hip joint is a common site for *referred pain*, that is, pathology and pain related symptoms to be referred to the hip, for example, from the knee. Referred pain means that pain due to a pathology at one site is referred, or felt, somewhere distant to the site of the pathology due to either shared nerves, or nerve roots. Therefore, examination of the hip joint should always be combined with examination of the vertebral column, pelvis, and also the knee joint.

3.3.2.3 Clinical Examination and Applications

At this point, it seems appropriate to discuss the clinical examination of the hip joint. The hip joint is a ball and socket synovial joint which is one of the most stable joints of the body. The deeply set acetabulum and the muscle bulk around the hip joint aids in stability enforcement. This joint is also remarkably flexible and mobile and allows for a wide range of movements including flexion, extension, abduction, adduction, and a combination of these movements referred to as circumduction.

However, like all areas in the body, the hip joint is also affected by age related change. As we grow older, the movement of the hip joint becomes more limited, with extension and internal rotation becoming reduced first, followed by abduction.

Pain from hip joint pathology is typically felt around the groin. However, it can also radiate to the front of the thigh down to the knee, or simply be felt only in the knee (i.e., referred pain). A patient who cannot easily localize knee pain, and especially if adolescent, could have a slipped femoral epiphysis.

The following list will provide detailed steps to follow in clinical examination of the hip joint.

3.3.2.4 Examination of the Hip Joint
1. As with all examinations a few simple steps should be followed.
2. Introduce yourself to the patient, ask their identity and always wash your hands according to local protocols.
3. Always obtain the patient's consent for any examination you undertake, no matter how trivial they may seem.
4. Two ways to approach the examination of the hip joint is to undertake examination in the erect and supine positions.

3.3.2.5 Erect Examination of the Hip Joint
1. Assess the patient standing and walking (i.e., gait).
2. Inspect for any scoliosis or tilting of the pelvis during standing and in movement.
3. Have the patient walk, unaided if possible, or with the necessary support to ensure they do not lose their balance.
4. Observe for the Trendelenburg gait, as described in the Superior Gluteal Nerve Injury section in Section 3.3.1 entitled Superior Gluteal Nerve.

5. Assess leg length – this can either be true (i.e., the leg is structurally shorter than the opposite side) or apparent (due to hip deformity).

6. A fixed abduction deformity results in apparent leg lengthening as the pelvis tilts down on that side bringing the leg parallel. The patient then adjusts their posture by flexing the knee on the affected side. This is to try to shorten the limb on that affected side. Or, the patient may raise the foot with, for example, in-soles, or more extreme footwear adjustments. However, the scoliosis still will not have changed.

7. A fixed adduction deformity results in apparent limb shortening. The pelvis is raised on the affected side.

8. With a fixed flexion deformity, the apparent limb shortening can be compensated by the patient increasing lumbar lordosis.

9. Single deformities like these just mentioned tend not to be common. The most common combination of hip joint positions is flexion, lateral rotation and adduction of the hip joint.

3.3.2.6 Examination of the Supine Patient for Hip Joint Pathology

3.3.2.6.1 Inspection

Look at the pelvis and area around the hip joint. Observe any obvious limb deformity or variation in position. Is there symmetry between left and right sides? Document what the initial appearance is with the patient supine. With the patient supine, make sure the pelvic brim is at right angles to the spine.

TIP!

It is essential to ensure that the pelvic brim is at right angles to the spine. Fixed flexion deformity will be partially or completely masked by the patient moving their pelvis forward and increasing the lordosis in the lumbar region. If a patient has a severe fixed flexion deformity at the hip joint, the knee will be flexed but it can also be passively extended. This differentiates it from a fixed flexion deformity of the knee due to pathology within the knee joint itself.

TIP!

A flexion deformity of the hip joint should differentiate between pathology at the hip, or from psoas spasm or irritation of the femoral nerve.

With psoas spasm, the patient will have the hip joint flexed; any attempt to straighten it is painful. Passive rotation, however, will not be limited and generally not painful. A number of causes of psoas spasm can include abdominal pathology (e.g., appendicitis, or perforated appendix), spinal abscess in the lumbar region or pelvic pathologies.

If there is hip joint pathology, then rotation of the joint and other hip related movements would be painful, or extremely painful to the patient.

3.3.2.6.2 Leg Length Measurement

Accurate measurement of limb length really should be done by radiological means. However, it may be apparent during clinical examination that there is limb shortening, for example, in fixed flexion deformity of the limb. Measurement of the limb length should be undertaken at the bedside by measuring from the anterior superior iliac spine to the medial malleolus. Both limbs must be in as comparable a position as each other.

3.3.2.6.3 Palpation

In terms of palpation, it can be difficult to detect anything at this site other than an obvious swelling. Examination should always include palpation over the greater trochanter to determine if a trochanteric bursitis is present. In addition, palpation can localize tenderness, and may indicate muscle spasm in and around the groin from, for example, an iliopsoas bursa.

3.3.2.6.4 Movement

With the iliac crest stabilized by the examiner's hand, assess hip flexion of each hip joint. The range of flexion can range from 0–120°.

- *Thomas's test* should be performed. This involves the examiner placing one hand under the patient's lumbar spine between the vertebral column at that position, and the examining couch. Then the hip should be flexed to straighten the lumbar spine. This will result in the spine pressing the examiner's hand, that is, the vertebral column will come closer to the examining couch. The examiner should ensure their own safety and not have their hand crushed by the patient, so exercise caution in the obese patient. Normally, the opposite limb will remain flat on the examining couch. If there is any pathology on the opposite side, it will result in a variable degree of leg raising.
- Abduction and adduction of each hip joint should be undertaken, and a record made of how much movement is possible in each. Typically a minimum 45° should be achieved during abduction, and 25° for adduction.
- Then, each limb should be rotated observing the movement of the foot through a 90° arc of motion.
- Rotation can also be measured with the hip and knee joints flexed, using the tibia as a reference point for medial rotation (30°) and lateral rotation (45°).

3.3.2.6.5 Hip Stability

Hip stability should be part of the normal examination of the hip joint. Typical clinical presentations involving hip instability include congenital dislocation of the hip, femoral neck fractures, or slipped epiphysis.

1. Congenital dislocation of the hip
   In children with a suspected congenital dislocation of the hip, this should be assessed and acted upon as soon as possible, and certainly before the child starts to walk. Typically, the limb on the affected side will have the following features:
   a. Limb shortening.
   b. External (lateral) rotation.
   c. Buttock folds will be asymmetrical.
   d. Reduced abduction. Typically in an infant 90° should be easy to undertake.

   Ortolani's test can be performed if these signs are present. This involves the following:
   a. The child should be supine.
   b. The examiner should place their forefingers over the greater trochanter on both sides. The thumbs should be placed on the medial aspect of the upper thigh.
   c. The hips should be flexed to 90° and then slow abduction on both sides at the same time.
   d. The greater trochanters should be moved forward, very gently. If there is a feeling of slipping when the femoral head goes into the acetabulum, it represents instability of the hip joint. Further investigation is merited according to local and national guidelines and protocols.
2. Femoral neck fractures
   Typically, this occurs especially in elderly women. The limb will appear shortened, externally (or laterally) rotated, and extended. This requires urgent investigation and surgical fixation if X-ray confirms fracture.
3. Slipped epiphysis
   This typically occurs during puberty when the upper femur epiphysis slips. The limb will appear laterally, or externally, rotated, shortened, and abducted.

3.3.3 Nerve to Quadratus Femoris

The nerve to quadratus femoris originates from the fourth and fifth lumbar vertebrae and the first sacral nerves. The nerve to quadratus femoris leaves the pelvis below the level of the piriformis, and then runs downward anterior to the sciatic nerve. It innervates the inferior gemellus, and then passes to gain entry to the anterior aspect of the quadratus femoris, but it also sends a branch to the hip joint.

The *quadratus femoris* muscle is a flat and small muscle, which extends from the superior lateral border of the ischial tuberosity. It extends laterally to attach to the quadrate tubercle on the femur. This is found at the intertrochanteric crest at the posterior part of the femur. The quadratus femoris performs two functions – external rotation of the hip joint and helps with hip adduction.

3.3.3.1 Clinical Examination

When undertaking any clinical history taking or examination, you should always do the following, and follow a logical and systematic format:

1. Introduce yourself to the patient.
2. Advise them of what position you hold, for example, student, specialty grade, consultant etc.
3. Your reason for consulting with them, or to find out why they have presented to you.
4. Always take a thorough and detailed history, which will be guided by the presenting signs and symptoms.
5. When examining the patient, always tell them what you will ask them to do, or what region of the body you will be examining, with specific instructions and ensure they give consent.

A detailed examination and history taking should be completed as described in Chapter 1. Examination of the nerve to quadratus femoris should involve a comprehensive examination of the hip joint as discussed in the previous subsection, and should make sure that the function of the nerve to quadratus femoris is assessed, that is, external rotation of the hip joint and hip adduction.

3.3.3.2 Clinical Applications

3.3.3.2.1 Quadratus Femoris Injury

Injury to the quadratus femoris is rare and few cases are presented in literature. O'Brien and Bui-Mansfield (2007) presented only seven cases and stated that a female predominance occurred in younger patients and the right side was more affected than the left side. This type of injury can be sport related, especially in tennis. It could be due to the control of hip rotation internally during serving the ball, and with marked eccentric contraction of the quadratus femoris, but a congenitally small distance between the ischial tuberosity proximally and the lesser trochanter has also been proposed (Kassarijian, 2008).

Incorrect diagnosis can happen due to the attachment site of the quadratus femoris being the same as the hamstrings. The patient typically presents with pain in the hip, groin, gluteal region or posterior compartment of the thigh proximally (O'Brien and Bui-Mansfield, 2007; Bano et al., 2010).

Trying to diagnose injury or tears to the quadratus femoris can be challenging. As the clinical presentation can mimic other conditions, visualizing the quadratus femoris is best by MRI scanning. Typically, a tear to the quadratus femoris can be seen with edema behind the lesser trochanter to the sagittal plane. It has to be kept in mind that in trying to differentiate it from a tear of the obturator externus muscle, the edema in that type of case would be found medial to the greater trochanter in the posterior thigh (O'Brien and Bui-Mansfield, 2007). However, due to the orientation of the quadratus femoris, and coronal sectioning on MRI, it can prove challenging to see.

Treatment of tears of the quadratus femoris tends to be conservative with stepwise physiotherapy to aid pain-free lateral rotation of the hip joint. Dependent on the symptoms and severity of the tear, surgical decompression may be needed (Bano et al., 2010).

3.3.4 Pudendal Nerve
The pudendal nerve (Figure 3.2) originates from the sacral nerves 2–4 (i.e., S2–S4) and provides innervation to the majority of the perineum. Within the pudendal nerve are sensory and motor fibers as well as postganglionic sympathetic fibers. After originating from S2 to S4, the pudendal nerve then passes through the greater sciatic foramen, below the level of the piriformis. It passes the back of the ischial spine, and is medial to the internal pudendal artery. It enters the perineum with the internal pudendal artery via the lesser sciatic foramen. In the lateral

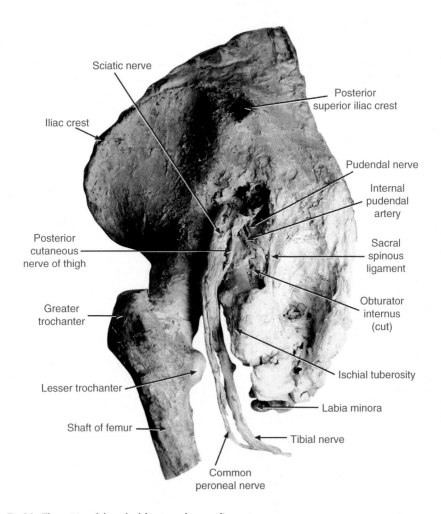

Fig. 3.2. The position of the pudendal nerve, and surrounding anatomy.

wall of the ischiorectal fossa, the pudendal nerve gives off three main branches:

1. Perineal nerve
2. Inferior rectal nerve
3. Dorsal nerve of the penis or clitoris

3.3.5 Perineal Nerve

In the pudendal canal, the perineal nerve gives off both deep and superficial branches. The deep branch of the perineal nerve pierces the medial wall of the pudendal canal and goes on to innervate the *levator ani* and the *external anal sphincter*. After this, the deep branch passes to supply the *bulbospongiosus, ischiocavernosus, superficial transverse perineal muscle*, and *bulb of the penis*.

The superficial branch of the perineal nerve divides into the posterior scrotal, or labial nerves, which are medial and lateral. Both of these divisions pierce the superficial and deep perineal fascia passing anterior with their corresponding arteries to pass to the *scrotum* or *labia majora*.

3.3.6 Inferior Rectal Nerve

The inferior rectal nerve can arise sometimes from the sacral plexus itself at the level of the third and fourth sacral vertebrae (S3–S4), as well as the pudendal nerve. It traverses the ischiorectal fossa via the medial wall of the pudendal canal. It goes on to supply the *external anal sphincter, skin of the anus*, and the *lining of the anal canal* as far superior as the *pectinate line*.

3.3.7 Dorsal Nerve of the Penis or Clitoris

This branch passes through the urogenital diaphragm providing innervation to the deep transverse perineal muscle and urethral sphincter. On piercing the inferior fascia of the urogenital diaphragm, it provides a branch to the *corpus cavernosum penis* or *clitoris*, and goes between two layers of suspensory ligaments of the penis or clitoris. It courses anteriorly on the dorsum of the penis or clitoris, innervating the *skin, prepuce*, and *glans*.

The branches of the pudendal nerve are summarized in Table 3.5.

3.3.8 Nerve to Obturator Internus

The nerve to obturator internus arises from the last lumbar vertebral level and the first and second sacral vertebrae, that is, L5–S2. On leaving the spinal cord at these levels, the nerve to obturator internus goes through the sciatic foramen inferior to the piriformis. The nerve to obturator internus gives off a small branch to the superior gemellus entering its deep aspect. It courses over the ischial spine on the lateral aspect of the internal pudendal vessels. It then goes through the lesser sciatic foramen and onto the obturator internus at its pelvic surface supplying it.

Indeed, there have also been shown to be variations in the nerve supply to the obturator internus. It has always been shown to enter

Table 3.5 The Branches of the Pudendal Nerve and the Specific Structures that are Innervated by them

Pudendal nerve branches		
Perineal nerve	Superficial branch	Scrotum/labia majora
	Deep branch	Levator ani, external anal sphincter, bulbospongiosus, ischiocavernosus, superficial transverse perineal muscle, Bulb of the penis
Inferior rectal nerve		External anal sphincter, skin of the anus, lining of the anal canal (as far superior as the pectinate line)
Dorsal nerve of penis/ clitoris		Corpus cavernosum penis or clitoris, skin, prepuce, and glans of penis or clitoris

the muscle on its medial aspect, spreading out anteriorly within the muscle itself. Also, it has been shown that some branches within the muscle also turn in a lateral and posterior direction to supply it (Aung et al., 2001).

The obturator internus arises from the internal surface of the obturator membrane and from the pelvis (see Figure 3.3). It extends from the greater sciatic notch to the pectineal line and inferiorly with the ischial and pubic margins of the obturator foramen. The tendon of obturator internus exits through the lesser sciatic foramen. It then passes anteriorly, and attaches onto the medial aspect of the greater trochanter in front of the attachment of the piriformis. A bursa is found between the tendon of obturator internus and the insertion on the lesser sciatic notch. The obturator internus is responsible for several functions namely:

- Lateral rotation of the extended hip
- Abduction of the extended hip
- Abduction of the femur when the hip is flexed

Hodges et al. (2014) also showed that the obturator internus, through electromyographic studies, were greatest in contracting during hip extension, then in lateral rotation of the hip, then on abduction. They showed that there was no or little contraction in other activities. This study provided greater insight into the function of the muscle as it has been previously difficult to assess due to difficulty in accessing such a deep muscle.

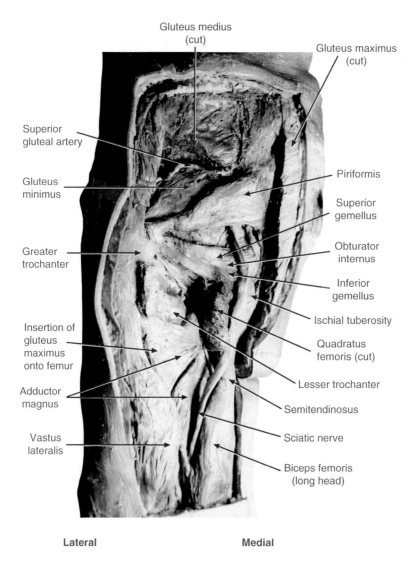

Gluteus medius (cut)

Gluteus maximus (cut)

Superior gluteal artery

Gluteus minimus

Greater trochanter

Insertion of gluteus maximus onto femur

Adductor magnus

Vastus lateralis

Piriformis

Superior gemellus

Obturator internus

Inferior gemellus

Ischial tuberosity

Quadratus femoris (cut)

Lesser trochanter

Semitendinosus

Sciatic nerve

Biceps femoris (long head)

Lateral   **Medial**

Fig. 3.3. The obturator internus and related anatomy.

3.3.9 Femoral Nerve

The femoral nerve (Figure 3.4) is from the lumbar plexus, and is the largest of its branches. It begins in the substance of the psoas major muscle emerging from its lateral border just inferior to the iliac crest. The femoral nerve then descends in the groove between the psoas

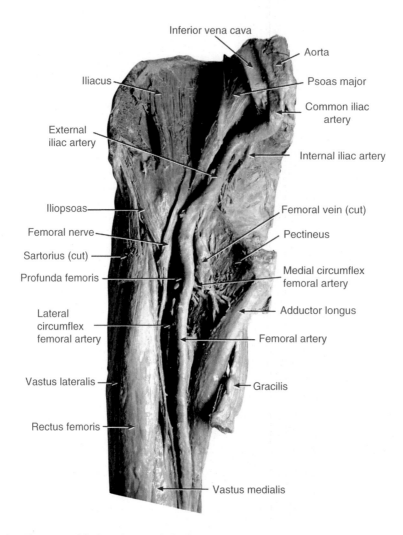

Fig. 3.4. The position of the femoral nerve and related structures.

major and the iliacus entering the thigh posterior to the inguinal ligament. It is located lateral to the femoral vessels.

TIP!

An easy way to remember the position of the femoral artery, nerve and vein at the proximal thigh is by the mnemonic *NAVY*. This stands for, from lateral to medial (at the pubic symphysis):

N: femoral *N*erve
A: femoral *A*rtery
V: femoral *V*ein
Y: *Y* fronts (type of male underwear centered at the pubic symphysis!)

The femoral nerve remains out with the femoral sheath as its origin from the lumbar plexus lies posterior to the plane of the fascia covering the psoas and iliacus musculature. In the thigh, the femoral nerve lies within the femoral triangle.

The femoral triangle is an anatomical triangle in the upper one-third of the anterior aspect of the thigh. It contains the femoral vessels and nerve, as well as lymph nodes. The boundaries of this triangle are as follows:

- Superior boundary: Inguinal ligament
- Lateral boundary: Sartorius (medial border)
- Medial boundary: Adductor longus (medial border)
- Roof: Fascia lata and cribriform fascia
- Floor: Iliopsoas, pectineus and adductor longus

Within the femoral triangle, the femoral artery is covered by the anterior aspect of the femoral sheath, the cribriform fascia superiorly, and the fascia lata inferiorly. Posteriorly, the femoral artery lies on the back of the femoral sheath and the psoas major, which separates it from the head of the femur. Inferior to this, it is separated from pectineus and adductor longus by the femoral vein. The femoral vein, which lies behind the artery in the lower part of the femoral triangle, winds to its medial side.

The *adductor canal*, also called the *subsartorial canal* is found within the middle one-third of the thigh. Through this canal passes the femoral vessels, saphenous nerve, and typically the nerve to vastus medialis. Laterally, it is bounded by the vastus medialis, medially by the adductor longus and also generally the adductor magnus.

Within the adductor canal, the femoral artery is covered anteriorly by the fascial roof of the adductor canal, sartorius and the subsartorial plexus. This canal is also called *Hunter's canal*, after *John Hunter*, a famous Scottish anatomist who first described it when ligating the femoral artery for a patient with a popliteal aneurysm. Popliteal aneurysms were extremely common in the late eighteenth century especially in "coach drivers, postilions, and others working in equestrian occupations in Georgian London" (Moore, 2005). In other words, the compression at the popliteal fossa and wearing of high-riding boots, resulted in the rise in rates of popliteal aneurysms during that era. John Hunter was pioneering at that time in the surgical correction of this problem, although not something used to treat this condition nowadays.

Returning to the femoral nerve, it has nerves, which arise at different regions, supplying a variety of motor and sensory structures/areas (summarized in Table 3.6). Within the abdomen, the femoral nerve can give off the lateral femoral cutaneous nerve. This supplies not only the iliacus muscle, but also sends small branches to the femoral artery. The innervation of pectineus arises from either within the abdomen, or posterior to the inguinal ligament or within the femoral triangle itself, so is variable. The nerve to pectineus also provides sensory supply to the hip joint. All of the other branches of the femoral nerve arise within the thigh.

The femoral nerve branches within the thigh can be classified as an anterior division (which has cutaneous and motor innervation) and the posterior division that again has motor and sensory nerves within it.

The anterior division, or group, is referred to as the intermediate and medial cutaneous nerves. The *intermediate femoral cutaneous* nerve

Table 3.6 The Branches of the Femoral Nerve, Where they Arise and Also Individual Branch Names and Final Structures they Innervate

Femoral nerve branches			
In abdomen	Lateral femoral cutaneous nerve		Iliacus, pectineus, femoral artery
	Nerve to pectineus		Pectineus, hip joint
Femoral triangle	Femoral nerve/intermediate cutaneous nerve		Sartorius
Thigh	Anterior cutaneous branches	Medial cutaneous nerves	Skin on medial side of thigh, subsartorial plexus, patellar plexus
		Intermediate cutaneous nerves	Sartorius, skin on anterior thigh, patellar plexus
	Posterior division	Muscular branches	Rectus femoris, vastus lateralis, knee joint (via filament from vastus lateralis branch), vastus intermedius (and knee joint), vastus medialis
		Saphenous nerve	Skin of medial side of leg and foot, knee joint, subsartorial plexus, patellar plexus

Notes: The *lateral femoral cutaneous nerve* can arise from the femoral nerve, but also can simply be an independent branch of the lumbar plexus, typically arising from the second and third lumbar vertebrae (L2–L3). The lateral femoral cutaneous nerve emerges from the lateral border of the psoas major, crossing the iliacus obliquely, passing into the thigh posterior to the inguinal ligament, near the anatomical landmark of the anterior superior iliac spine. It has multiple potential relations to sartorius – either passing behind, in front of, or even through this muscle.

supplies the sartorius and then passes through the fascia to innervate the anterior thigh and also contributes to the patellar plexus. The *medial femoral cutaneous nerve* passes superficial to the femoral vessels at the apex of the femoral triangle (inferior aspect) and innervates the medial aspect of the thigh and also provides branches to the patellar and subsartorial plexus.

The posterior division branches to form the saphenous nerve and also provides muscular branches. The saphenous nerve is the terminal branch of the femoral nerve. It continues with the femoral vessels through both the femoral and subsartorial triangles. In the femoral canal, it passes from the lateral to the medial side of the femoral artery. Then, with the saphenous branch of the descending genicular artery, passes to the skin betwcen the gracilis and the sartorius. The saphenous nerve runs in the leg with the great, or long, saphenous vein, and innervates the skin of the medial side of the leg and also the foot. It also gives off some branches to the subsartorial and patellar plexuses. These branches and the structures they innervate are summarized in Table 3.6.

Another important point to note is the *genitofemoral nerve*. This nerve arises from the first and second lumbar vertebrae (L1–L2). After leaving the vertebral column, it then passes over the front surface of the psoas major muscle, then passes obliquely behind the ureter, and then divides into its genital and femoral branches deep to the inguinal ligament, on the lateral aspect of the femoral artery.

The genital branch supplies the cremaster muscle and provides sensory innervation to the anterolateral scrotum. In females, the genital branch of the genitofemoral nerve supplies the *labia majora* and the mons pubis. In addition, the femoral branch supplies the skin superficial to the femoral triangle, lateral to the territory supplied by the ilioinguinal nerve.

3.3.9.1 Clinical Applications
3.3.9.1.1 Genitofemoral Nerve Block
As mentioned previously, the genitofemoral nerve originates from L1 to L2. It has been reported that postoperative pain after inguinal hernia repair was problematic and occurred frequently (Peng and Tumber, 2008; Campos et al., 2009). Blockade of the genitofemoral nerve during inguinal hernia repair and selective blocks of the genitofemoral and ilioinguinal nerves are sufficient in procedures like testicular

Table 3.7 What to Ask the Patient to do, when Testing the Specific Nerve Roots of the Femoral Nerve, and What Muscles are Being Tested

Nerve Root	Muscle	Ask the Patient to...
L1–L3	Iliopsoas	Flex their hip with the knee flexed and the lower leg supported. The patient should be supine
L2–L3	Sartorius	Flexion of the knee with the hip joint in external rotation
L2–L4	Quadriceps femoris	Extension of the knee against resistance by the examiner

biopsies. This provides adequate analgesia and anesthesia for these procedures, reducing postoperative pain (Al-Alami et al., 2013).

3.3.9.2 Clinical Examination

When undertaking any clinical history taking or examination, you should always do the following, and follow a logical and systematic format:

1. Introduce yourself to the patient.
2. Advise them of what position you hold, for example, student, specialty grade, consultant etc.
3. Your reason for consulting with them, or to find out why they have presented to you.
4. Always take a thorough and detailed history, which will be guided by the presenting signs and symptoms.
5. When examining the patient, always tell them what you will ask them to do, or what region of the body you will be examining, with specific instructions and ensure they give consent.

A detailed examination and history taking should be completed as described in Chapter 1.

The above Table 3.7 summarizes the nerve roots, muscle supplied and relevant clinical test to ask the patient to perform in assessing the femoral nerve.

3.3.10 Sciatic Nerve

The sciatic nerve (Figure 3.5) is the largest nerve in the body. It originates from the lumbosacral plexus from the fourth lumbar through to the third sacral segments (i.e., L4–S3). Typically, the sciatic nerve leaves the pelvis through the greater sciatic foramen, usually inferior to

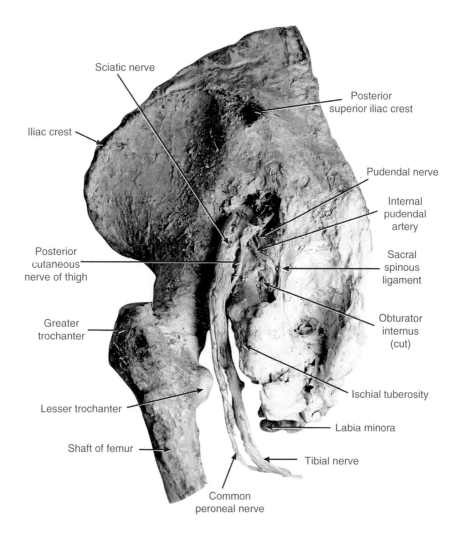

Sciatic nerve

Posterior
superior iliac crest

Iliac crest

Pudendal nerve

Internal
pudendal
artery

Posterior
cutaneous
nerve of thigh

Sacral
spinous
ligament

Greater
trochanter

Obturator
internus
(cut)

Lesser trochanter

Ischial tuberosity

Labia minora

Shaft of femur

Tibial nerve

Common
peroneal nerve

Fig. 3.5. The path of the sciatic nerve and its two major branches – the common peroneal and tibial nerves.

the piriformis to enter the gluteal region. However, anatomical variation in this pattern typically exists in the relationship between the sciatic nerve and the piriformis muscle. An extensive review of over 6000 cadaver dissections in this region by Smoll (2010) highlighted that almost one-fifth of people may an anatomical anomaly in the relationship between the piriformis muscle and the sciatic nerve, not fitting into this classical relationship.

Although the sciatic nerve passes through the gluteal region, it does not supply structures there. The sciatic nerve, after generally passing inferior to piriformis, descends under the gluteus maximus between the greater trochanter and the ischial tuberosity. The sciatic nerve is the most lateral structure leaving the greater sciatic foramen, with the inferior gluteal vessels and nerve, pudendal nerve and internal pudendal vessels medial to the nerve entering the thigh posterior to the adductor magnus. The inferior gluteal artery supplies its own branch to the sciatic nerve aptly named the *artery to the sciatic nerve*. The sciatic nerve lies on the ischium and courses behind the obturator internus, quadratus femoris, and the adductor magnus. The sciatic nerve courses through the posterior compartment of the thigh deep to biceps femoris.

The sciatic nerve supplies both articular and muscular structures before it splits into its final common peroneal and tibial nerve branches.

1. *Articular*: The articular branches of the sciatic nerve proximally pass to supply the *hip joint*. These branches perforate the posterior surface of the capsule, and occasionally are derived from the sacral plexus itself.
2. *Muscular*: The muscular branches of the sciatic nerve innervate the hamstring muscles, that is, *biceps femoris, semimembranosus* and *semitendinosus*, as well the *ischial head of the adductor magnus*.

The sciatic nerve is a combination of the common peroneal nerve and the tibial nerve (Figure 3.6), as mentioned previously. The tibial nerve is from the anterior divisions of the anterior rami and the common peroneal nerve is from the posterior divisions of the anterior rami.

Although, in general, it is thought that the sciatic nerve divides into the tibial and common peroneal nerves at the upper region of the popliteal fossa, variations in this branching pattern do exist.

Broadly, there are four types of branching pattern of the sciatic nerve that occur. A Type I branching pattern of the sciatic nerve is where the common peroneal nerve passes through the main substance of the piriformis muscle, with the tibial nerve emerging from its inferior surface. Type I branching is found in approximately 10% of

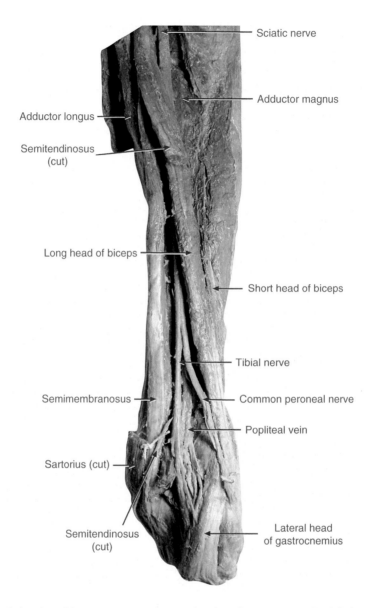

Fig. 3.6. The branching of the sciatic nerve into its two main branches – the common peroneal and tibial nerves.

people. Type II branching of the sciatic nerve (in approximately 2–3% of people; Machado et al., 2003) is where the common peroneal nerve passes over the superior border of the piriformis, with the tibial nerve passing below the inferior aspect of the muscle. Type III branching of the sciatic nerve is also less common, existing in approximately 2% of people. In Type III variation, the sciatic nerve passes through the piriformis and does not divide at all (Anson and McVay, 1971; Machado et al., 2003). In Type IV pattern of the sciatic nerve, the nerve passes above piriformis without dividing (Williams et al., 1989). However, many variations of this do exist (Vloka et al., 2001; Babinski et al., 2003).

3.3.10.1 Clinical Examination

Clinical examination of the sciatic nerve is by both active and passive movements to demonstrate the function, and any pathology involving the sciatic nerve, and related problems with it (Miller, 2007).

1. *Supine examination of sciatic nerve*
   a. Straight leg raising.
   b. As always, introduce yourself to the patient, ascertain their identity, take a full history and obtain consent when undertaking any examination.
   c. When examining the sciatic nerve for any related pathology, the straight leg test is simple and effective in ascertaining the problem.
   d. With the patient in the supine position, and with the knee flexed, check passive hip flexion is normal on both sides.
   e. It is best to examine the unaffected side first, if possible, to minimize pain to the patient.
   f. Then, with the knee extended, the clinician (after informing the patient what you will be doing) raises the leg on the unaffected side. Ensure that the knee is prevented from flexion with the other examining hand.
   g. Observe the range of motion.
   h. Then, this is repeated on the opposite side, that is, raising of the patient's leg with one hand from the ankle, with the other hand making sure the knee stays in the extended position.
   i. Make sure you do not cause pain to the patient though, and stop the examination if the patient is not able to undertake the examination.
   j. Make sure you advise the patient to tell you where exactly they feel pain and/or paresthesia, and where it radiates to.

   k. When the limit of hip flexion is reached, with the knee extended, gently dorsiflex the ankle joint, which places further stress on the sciatic nerve. This is called *Bragaard's test.*

2. *Bowstring sign*
   a. Undertake the straight leg raising again with the patient supine and at the limit for the patient, bend (flex) the knee, which will reduce the tension on the sciatic nerve.
   b. Then, keep flexing the hip joint to approximately 90°, or what the patient can maximally do.
   c. GENTLY, extend the knee again until pain starts to become apparent. This is called *Lasègue's sign.*

3. *Sitting test*
   a. With the patient lying down, ask the patient to sit up from the lying position.
   b. Only if sciatic nerve irritation is NOT present, will a patient be able to sit up with legs flat on the examining bed.

4. *Flip test*
   a. With the patient sitting at the edge of the examining bed, with the knees and hips flexed, test for knee jerk reflexes.
   b. Then extend the knee to start examining the ankle jerk reflex.
   c. If genuine sciatic nerve pathology is present, the patient will "flip" back to relieve the tension.

5. *Neri's bowing sign*
   a. With the patient standing, ask them to bend forward as if to touch their feet.
   b. If the patient shows flexion of the knee on the affected sign it is a positive sign of potential sciatic nerve irritation, or indeed strain of the sacroiliac joint or radicular pain. So, it is not as specific as the others previously mentioned.

Additional signs of pathology of the sciatic nerve, or irritation at the L5/S1 nerve roots, will include weakness of the following:

1. Dorsiflexion of the great toe
2. Ankle dorsiflexion
3. Ankle reflex

The patient may also have numbness and/or paresthesia in the first interdigital cleft, lateral border of the foot and sole and lateral aspect of the calf.

Table 3.8 The Nerve Roots of the Sciatic Nerve, the Muscles They Supply, and What to Ask the Patient to Elicit Functioning (or Otherwise) of the Nerve Roots and Muscles

Nerve Root	Muscle	Ask the Patient to...
L4–S2	Hamstrings	Flexion of the knee against resistance
L4–L5	Tibialis posterior	Inversion of the foot in plantar flexion
L4–L5	Tibialis anterior	Dorsiflexion of the ankle joint
L5–S1	Extensor digitorum longus	Dorsiflexion of the toes against resistance
L5–S1	Extensor hallucis longus	Dorsiflexion of the great toe against resistance
L5–S1	Peroneus (fibularis) longus and brevis	Eversion of the foot against resistance
S1	Extensor digitorum brevis	Dorsiflexion of the great toe
S1–S2	Gastrocnemius	Plantar flexion of the ankle joint
S1–S2	Flexor digitorum longus	Flexion of the terminal joints of the toes
S1–S2	Small muscles of the foot	Bring the sole of the foot into a cup

In addition to this, Table 3.8 can be used as a summary as to the specific nerve roots of the sciatic nerve, what muscles are being tested, and what to ask the patient to do in determining the functions of each category.

3.3.10.2 Clinical Applications

3.3.10.2.1 Sciatica

Sciatica shows considerable variance in the incidence reported in the literature, ranging from 1.6% to a staggering 43% from a general working population. Sciatica is a symptom, which is found in the lower back and also the hip. The pain found in sciatica typically radiates down the back of the thigh, leg and can reach the foot. Most often, it is caused by a slipped disc, that is, herniation of a lumbar intervertebral disc, which compresses either the L5 or S1 nerve. Approximately 95% of lumbar intervertebral discs which compress the nerve root occurs at L4/5 or L5/S1 hence affecting the L5 or S1 nerve roots respectively (Moore and Dalley, 2006).

As we age, the space between the intervertebral foramina reduces and the nerves can increase in size. This coupled with new osteophyte deposition at the exit/entry point of the nerves also amplify this problem, which can compress the nerve roots of the sciatic nerve, giving the classical presentation. The examination of suspected sciatica has been mentioned previously.

Treatment of sciatica can range from conservative measures through to surgery, and involves the full spectrum in between. It has been quoted by the National Health Service in the United Kingdom that sciatica generally improves on its own after approximately 6 weeks (NHS, 2015). Gentle exercise is recommended to help alleviate signs and symptoms like walking, gentle stretch exercises, and nonweight bearing exercises, for example, gentle swimming, though not too strenuous if it is unbearably painful. Over the counter (OTC) medications for pain relief can be ample, but it may be that stronger painkillers are required, spinal injections (of steroids or local anesthetic agents), or in severe cases, surgery to undertake discectomy, laminectomy or fusion of the vertebrae (NHS, 2015).

As well as sciatica occurring through increasing age, it may also result from sport related injury where there is hyperflexion of the cervical region, for example, in American football or rugby during a tackle. This type of high impact tends to affect the posterior component of the disc, but not resulting in a fracture of the vertebral body in the cervical region. In these cases, the nerve affected is at that level of injury, and not the nerve below, as previously described. The reason for this is that the cervical spinal nerves arise superiorly to that vertebral level. When the rupture of the disc happens in the cervical region, it results in neck, shoulder, arm, and hand pain.

3.3.11 Obturator Nerve

The obturator nerve arises from the lumbosacral plexus, specifically from the third and fourth lumbar vertebrae, but occasionally from L2 and/or L5 too. The obturator nerve then descends through the substance of the psoas major, a major hip flexor, emerging from the medial margin of the psoas major, at the level of the pelvic inlet posterior to the common iliac vessels. It then accompanies the superior lying obturator vessels passing to the obturator groove on the lower aspect of the superior ramus of the pubis, where it then divides into an anterior and posterior branch. These branches then pass inferiorly to enter the obturator foramen on their way to the thigh, where the adductor brevis lies between the branches. Ultrasonic imaging of the obturator nerve and its related branches described, has shown it to be the flattest of all the peripheral nerves that can be imaged (Soong et al., 2007).

3.3.11.1 Anterior Branch

The anterior branch of the obturator nerve lies anterior to the obturator externus and the adductor brevis, and posterior to the pectineus and the adductor longus. This branch terminates along the medial border of the adductor longus as a filament that supplies the femoral artery. The anterior branch of the obturator nerve supplies the *adductor longus, adductor brevis, gracilis*, and occasionally the *pectineus*. There can also be a branch (variable) that forms part of the *subsartorial plexus* supplying the skin on the medial side of the thigh and leg. Sometimes, a small filament may also be found which passes to the knee joint.

3.3.11.2 Posterior Branch

The posterior branch of the obturator nerve (Figure 3.7) pierces through the obturator externus. It then passes inferiorly and anterior to the adductor magnus and posterior to the adductor brevis. It terminates by passing into the adductor magnus, occasionally with the femoral artery, and then descends on the popliteal artery, piercing the oblique popliteal ligament supplying the knee joint. It also gives muscular branches to the *obturator externus, adductor magnus*, and occasionally *adductor brevis*.

3.3.11.3 Accessory Obturator Nerve

The accessory nerve arises from either the third and fourth lumbar vertebrae or the second and third lumbar vertebrae. It is inconstant, but when present, emerges from the medial margin of the psoas major and enters the thigh anterior to the pubis. It communicates with the anterior branch of the obturator nerve, supplying branches to the *pectineus*, but also the *hip joint*. It may also supply the *adductor longus*, but its presence is not common (Akkaya et al., 2008).

3.3.11.3.1 Adductor Longus

This muscle attaches from the body of the pubis below the pubic crest proximally, passing to the linea aspera of the femur, approximately in its mid-third portion.

3.3.11.3.2 Adductor Magnus

The adductor magnus has two parts to it – the adductor portion, and the hamstring part. The adductor portion attaches proximally to the

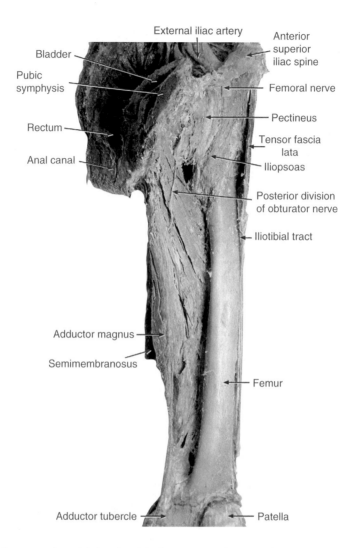

Bladder

Pubic
symphysis

Rectum

Anal canal

External iliac artery

Anterior
superior
iliac spine

Femoral nerve

Pectineus

Tensor fascia
lata

Iliopsoas

Posterior division
of obturator nerve

Iliotibial tract

Adductor magnus

Semimembranosus

Femur

Adductor tubercle

Patella

Fig. 3.7. The position of one of the branches of the obturator nerve – the posterior branch.

inferior ramus of the pubis, and also the ramus of the ischium. The hamstring part arises from the ischial tuberosity.

The distal attachments vary depending on if it is the adductor or hamstring portion. The adductor part attaches distally to the linea aspera, the gluteal tuberosity on the femur and the medial supracondylar line on the distal femur. The hamstring portion attaches onto the femur at the adductor tubercle on the inferomedial portion of the bone.

3.3.11.3.3 Adductor Brevis
The adductor brevis attaches proximally to the body and also the inferior ramus of the pubis passing to the pectineal line and the linea aspera on the femur, but only its proximal portion.

3.3.11.3.4 Obturator Externus
The obturator externus attaches proximally to the obturator foramen margins and membrane and passes to attach distally onto the femur, at the trochanteric fossa. The obturator externus is responsible for adduction of the thigh and also aids medial rotation of the thigh.

3.3.11.3.5 Pectineus
The pectineus muscle attaches proximally to the pubis (superior ramus) passing distally to attach to the pectineal line on the upper medial aspect of the femur. Pectineus is responsible for adduction and flexion of the thigh, and helps somewhat in rotation of the thigh medially.

3.3.11.3.6 Gracilis
This long thin strap muscle attaches proximally to the inferior ramus of the pubis and its body passing to its distal attachment site at the upper medial portion on the tibia. Again, gracilis is responsible for adduction of the thigh, flexion of the leg, but also medial rotation of the thigh. The name gracilis comes from its Latin meaning of slender.

3.3.11.4 Clinical Applications
3.3.11.4.1 Actions of the Adductor Muscles
In the anatomical position, as the names of these muscles suggest (adductor longus, magnus, and brevis), they adduct the thigh, or allow it to be brought medially toward the center of the body. All of the adductors adduct the thigh but also the adductor brevis has a role in flexion of the thigh, and the adductor magnus has two additional roles. The adductor portion not only adducts the thigh but also flexes the thigh. The hamstring portion is responsible for extension of the thigh.

The typical sport where the adductors work most notably is in horse riding, where the adductors have to work to maintain the rider on the saddle.

Day-to-day, the adductors work typically as follows:

1. A stabilizer of the body when standing on both feet

2. Correction of a lateral swing of the upper body
3. Stabilizing the body when transferring weight from one foot to the other

Although it may be thought of as the adductors being essential to this adduction process, it has been shown that you can have up to a 70% reduction in their function with only slight or moderate impairment of function (Markhede and Stener, 1981).

Clinical testing of the adductor muscle group is done as follows:

1. Have the patient supine.
2. The knee should be straight.
3. The patient should then adduct the thigh against resistance placed by the examiner.
4. If normal, the adductor longus and gracilis can easily be palpated.

One other clinical test is referred to as the *slump test*. The *slump test* is a nonspecific test for nerve problems in a patient. It however extends from the head to the foot, so is not ideal in isolating where the lesion or problem is.

Testing of the obturator nerve can involve some modifications to the *slump test* to show if there are any localized problems to the groin. With the patient sitting, with legs hanging off, for example, the bed, the patient puts their hands clasped behind their back, flexes their neck and the thigh on the side where the problem is anticipated is abducted slowly, when the patient raises their head. This is seen as a mechanoreceptive test referred to as the *slump knee bend test* (Magee et al., 2009). It is, however, not a reliable test to perform.

3.3.11.4.2 Obturator Nerve Blockade

The obturator nerve can be anesthetized, or blocked, for pain control after knee surgery, for example, total knee replacement. This would typically involve a three nerve block – femoral, obturator and lateral cutaneous nerves, with variations in the location of the injections to give adequate analgesia (Choquet et al., 2005; Macalou et al., 2004).

In addition, obturator nerve blocks can be undertaken during transurethral resection surgery as the obturator nerve passes close to the bladder wall (Thallaj and Rabah, 2011). It is important to anesthetize this nerve to prevent adductor contraction during this type of surgery.

3.3.12 Tibial Nerve

The tibial nerve originates from the fourth and fifth lumbar vertebrae and the first three sacral vertebrae (L4–S3). The tibial nerve is incorporated into the sciatic nerve in the gluteal area and thigh, and descends as a separate nerve in the popliteal fossa. The tibial nerve is the larger terminal branch of the sciatic nerve. It arises in the inferior one-third of the thigh. It then sits on the popliteus, under the cover of the gastrocnemius. At the inferior aspect of the popliteus, it then courses deep to the fibrous arch of soleus reaching the back of the leg. It then courses through the leg by passing on the tibialis posterior and the flexor digitorum longus, coming to lie on the tibia. Inferiorly, it is more superficial. Initially, the tibial nerve is medial to the posterior tibial artery and then posteriorly, it crosses this vessel to lie on its lateral side (Figures 3.8 and 3.9). The tibial nerve terminates under the cover of the flexor retinaculum dividing into the lateral and medial plantar nerves (Figure 3.10).

TIP!

In terms of surface anatomy, the course of the tibial nerve can be visualized as if a line was drawn from about the level of the tuberosity of the tibia, inferiorly to a point mid-way between the medial malleolus and the heel.

Branches of the tibial nerve:

When incorporated in the sciatic nerve, the tibial nerve gives branches to the following muscles:

1. Semitendinosus
2. Semimembranosus
3. Long head of biceps
4. Adductor magnus

In the popliteal fossa, branches are given off at this point to the knee joint. As the tibial nerve descends down past the popliteal fossa, the following branches then are given off:

1. *Muscular branches* to
   a. gastrocnemius,
   b. soleus,

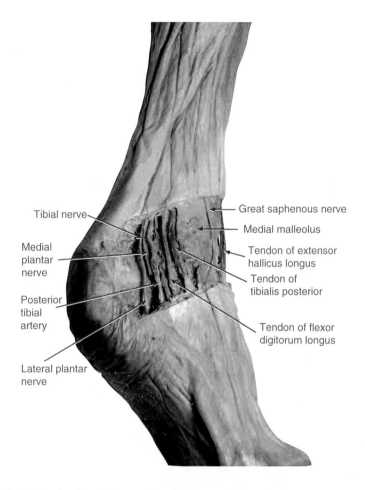

Tibial nerve

Medial plantar nerve

Posterior tibial artery

Lateral plantar nerve

Great saphenous nerve

Medial malleolus

Tendon of extensor hallicus longus

Tendon of tibialis posterior

Tendon of flexor digitorum longus

Fig. 3.8. The distal portion of the tibial nerve and its relations to vasculature and tendons.

    c. plantaris,
    d. tibialis posterior,
    e. popliteus.
      Note: The twig to popliteus muscle turns around the inferior border of the muscle to reach round to its anterior surface, accompanied by the posterior recurrent branch of the anterior tibial artery.
2. *Interosseous* nerve of the leg: This branch of the tibial nerve passes distally on the interosseous membrane and reaches the level of the tibiofibular syndesmosis.

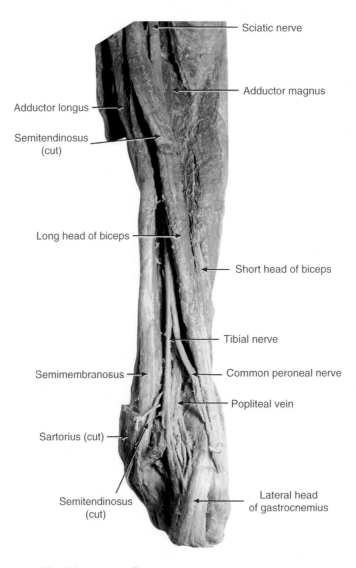

Fig. 3.9. The course of the tibial nerve proximally.

3. *Cutaneous branches*:
   a. *Medial sural cutaneous nerve*: This cutaneous branch of the tibial nerve descends between the two heads of gastrocnemius. It is usually joined by the peroneal communicating branch of the common peroneal nerve forming the sural nerve.

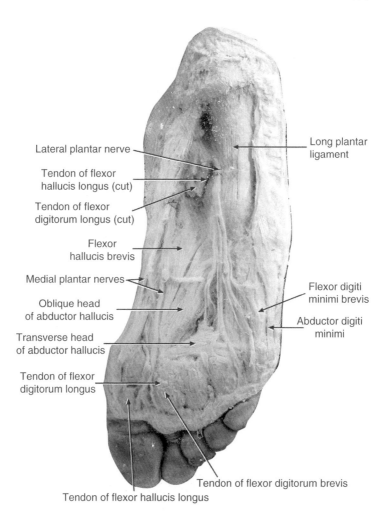

Lateral plantar nerve

Tendon of flexor
hallucis longus (cut)

Tendon of flexor
digitorum longus (cut)

Flexor
hallucis brevis

Medial plantar nerves

Oblique head
of abductor hallucis

Transverse head
of abductor hallucis

Tendon of flexor
digitorum longus

Long plantar
ligament

Flexor digiti
minimi brevis

Abductor digiti
minimi

Tendon of flexor digitorum brevis

Tendon of flexor hallucis longus

Fig. 3.10. The medial plantar nerves on the sole of the foot.

b. *Sural nerve*: This nerve passes between the heads of
gastrocnemius passing deep to the fascia in the superior part
of the leg posteriorly. It is joined by the sural communicating
branch from the common peroneal nerve. The sural nerve
then passes inferiorly to lie on the tendo calcaneus then, with
the small saphenous vein, passes to the posterior aspect of the
lateral malleolus. It gives off *lateral calcaneal branches* to
the skin on the posterior aspect of the leg and lateral aspect

of the foot and heel. It also gives off branches to the ankle joint and nearby tarsal joints. The forward continuation of this nerve toward the little toe communicates with the superficial peroneal nerve and is referred to as the *lateral dorsal cutaneous nerve*. This supplies the lateral aspect of the little toe and digital joints adjacent to this.

4. *Medial plantar nerve*: The medial plantar nerve (Figure 3.10) is the larger of the terminal branches of the tibial nerve, arising from underneath the flexor retinaculum. Initially the medial plantar nerve lies deep to the abductor hallucis, and then it passes anterior in the sole coming to lie between the abductor hallucis and the flexor digitorum brevis. The nerve supply of the medial plantar nerve is highlighted in Table 3.9.

5. *Lateral plantar nerve*: The lateral plantar nerve (Figure 3.11) is the other terminal branch of the tibial nerve. It arises from below the flexor retinaculum and passes anterior, deep to the abductor hallucis and flexor digitorum brevis. On its course, the lateral plantar nerve is accompanied by the lateral plantar artery (from the posterior tibial artery), which is on the lateral aspect of it. When it arrives at the base of the fifth metatarsal, it then subdivides into a deep and superficial branch. The individual branches are listed in Table 3.10.

3.3.12.1 Clinical Applications

As a simple overview, the tibial nerve supplies the following:

1. *Plantar flexors of the foot*

   a. Gastrocnemius
   b. Soleus
   c. Popliteus
2. *Invertor of the foot*

   a. Tibialis posterior

Table 3.9 The Structures Supplied by the Medial Plantar Nerve	
Medial plantar nerve	
Muscular branches	Flexor hallucis brevis, flexor digitorum brevis, abductor hallucis, 1st lumbrical
Cutaneous innervation	Medial three and one-half toes, nail beds and tips (dorsal)

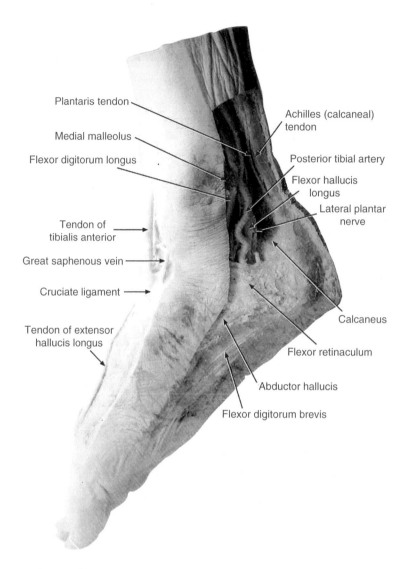

Plantaris tendon

Medial malleolus

Flexor digitorum longus

Tendon of
tibialis anterior

Great saphenous vein

Cruciate ligament

Tendon of extensor
hallucis longus

Achilles (calcaneal)
tendon

Posterior tibial artery

Flexor hallucis
longus

Lateral plantar
nerve

Calcaneus

Flexor retinaculum

Abductor hallucis

Flexor digitorum brevis

Fig. 3.11. The origin of the lateral plantar nerve.

3. *Flexors of the toes*

    a. Flexor hallucis longus
    b. Flexor hallucis brevis
    c. Flexor digitorum brevis
    d. Flexor digiti minimi

Table 3.10 The Branches of the Lateral Plantar Nerve, and the Related Subdivisions and What They Supply in Terms of Motor and Sensory Innervation

Branch Origin	Innervation
Main trunk	*Motor*: quadratus plantae, abductor digiti minimi; *sensory*: cutaneous innervation of the lateral part of the sole
Superficial branch	*Motor*: flexor digiti minimi brevis, interosseous muscles of fourth intermetatarsal space; *cutaneous*: sides of the lateral one and one-half toes, nail beds and tips of toes (dorsal)
Deep branch	*Motor*: adductor hallucis, 2nd–4th lumbricals, all interossei (except those in the fourth intermetatarsal space), articular twigs

3.3.12.2 Clinical Examination

1. As with all examinations a few simple steps should be followed.
2. Introduce yourself to the patient, ask their identity and always wash your hands according to local protocols.
3. Always obtain the patient's consent for any examination you undertake, no matter how trivial they may seem.
4. When assessing lower limb muscles and nerves, always keep in mind muscle tone, power, functional ability of the area being examined, reflexes and sensation (or abnormalities with them).
5. Testing plantar flexion of the toes, and the "big" toe assesses the tibial nerve. Ask the patient when lying flat, to push their toes downward.

TIP!

When assessing muscle activity and movement, always do it against resistance. In the case of plantar flexion a gentle pressing of the toes in a dorsiflexion direction when the patient plantar flexes allows you to test the strength of the muscle contraction. Always compare both left and right hand sides to assess muscle strength and tone.

---

6. For completeness, an assessment of the *ankle jerk reflex* can be undertaken. This assesses nerve roots S1 and S2.
7. Assessment of the ankle jerk reflex is undertaken with the patient lying flat. The foot, ankle and calf should be exposed on both sides as it is imperative to examine both left and right ankle jerks.
8. The leg should be resting on the examining couch.
9. Laterally rotate the hip and pull the foot SLIGHTLY into the dorsiflexed position.

10. Tap the calcaneal (Achilles) tendon with the tendon hammer lightly.
11. Observe the calf musculature for contraction. A positive response would elicit brisk plantar flexion of the foot.
12. A reduced or absent ankle jerk reflex can indicate pathology of the tibial and/or sciatic nerve.
13. Assessment of the sensory distribution of the tibial nerve should also be undertaken, that is, assessment of the *sural nerve* (*posterolateral region of the leg, down to the lateral aspect of the foot and fifth toe*), *medial calcaneal branches* (supplying *the heel and medial aspect of the sole of the foot posteriorly*), and the *plantar nerves* (*supplying the rest of the foot and toes*).
14. Light touch examination (using a cotton wool ball) and two-point discrimination should be undertaken after informing the patient what you will be doing, and with their eyes closed during the examination (Rea, 2015).

3.3.12.2.1 Tibial Palsy

A tibial palsy can be due to a variety of reasons, with the most common being knee lacerations or tibial or ankle fractures (Russell, 2006). Tibial palsy will present with weakness or paralysis of the muscles it supplies, that is, plantar flexors (soleus and gastrocnemius), inverting muscles (tibialis posterior), toe flexors (flexor hallucis longus, flexor hallucis brevis, flexor digitorum brevis, and flexor digiti minimi), and the intrinsic muscles of the feet. Dependent on the signs that the patient exhibits will help localize the anatomical position of the pathology. Refer to Table 3.11 for a summary of the location of a lesion to the tibial nerve and the resulting presentations typically found.

3.3.12.2.2 Tibial Nerve Entrapment

As the tibial nerve leaves the posterior compartment of the leg, it passes between the medial malleolus and the calcaneus deep to the flexor retinaculum. This area where the tibial nerve passes deep to the flexor retinaculum is referred to as the *tarsal tunnel*. As well as the tibial nerve being within this tunnel, a variety of other anatomical structures are held tightly in position. From anterior to posterior, the mnemonic Tom, Dick, and Harry can act as an aide-memoire for the structures in the tarsal tunnel, as follows:

Table 3.11 The Typical Clinical Presentations of Pathologies Involving the Tibial Nerve, Dependent on the Anatomical Site

Site of Pathology	Clinical Presentation
Popliteal fossa (or proximal)	Weakness of gastrocnemius and soleus, weakness/paralysis of all other muscles distal supplied by the tibial nerve
Deep posterior compartment or mid-leg (or inferior)	Sparing of the tibialis posterior, weakness/paralysis of all other muscles distal supplied by the tibial nerve
Lower third of leg	Sparing of flexor hallucis longus and flexor digitorum, weakness/paralysis of all other muscles distal supplied by the tibial nerve

TIP!

Tom: *T*ibilais posterior
Dick: flexor *D*igitorum longus
*AN*d: posterior tibial *A*rtery (and vein) and tibial *N*erve
*H*arry: flexor *H*allucis longus

See Figure 3.8 in identifying these structures.

Tarsal tunnel syndrome is the most common nerve entrapment at the ankle joint, and is similar to carpal tunnel syndrome, found in the wrist. It is classified as an entrapment, or compression neuropathy affecting the posterior tibial nerve, or indeed its more terminal branches, that is, plantar or calcaneal nerves.

Tarsal tunnel syndrome is uncommon, but of those cases it is identified in, the vast majority has a specific reason for the pathology (Havel et al., 1988). Factors causing tarsal tunnel syndrome are varied and can be divided into intrinsic or extrinsic factors. Those intrinsic factors include space occupying lesions, osteophytes or hypertrophic retinacular tissue. Extrinsic reasons include trauma to the ankle joint, restrictive footwear, systemic disease, or pregnancy (Ahmad et al., 2012).

Typically patients present with pain at the site of the tarsal tunnel, at the medial malleolus, and the pain radiates to the plantar aspect of the foot. The pain tends to be worst when they are going to bed; it can wake the patients from their sleep. In addition, when weight bearing, the intrinsic muscles of the foot are affected and are painful.

On examination, there is a positive Tinel's sign. Tinel's sign, named after Jules Tinel, a French neurologist, is a sign where, on lightly tapping a nerve, paresthesia occurs over the distribution of that nerve.

Formal investigation of the cause of tarsal tunnel syndrome, and to plan treatment based on the findings, can range from X-ray, MRI (Reade et al., 2001), and more recently ultrasound (Nagaoka and Matsuzaki, 2005).

> **TIP!**
>
> When trying to demonstrate a positive Tinel's sign, DO NOT tap the area where the nerve is expected as it can sometimes be incredibly painful for the patient.

3.3.12.2.3 Tibial Nerve Stimulation

One area that has been receiving increased attention is *tibial nerve stimulation*. In the United States alone, it has been quoted that about 34 million adults are affected by a condition called *overactive bladder* (OAB; Stewart et al., 2003). Patients with OAB typically present with frequency of urination throughout the day, nocturia, and with or without urgency incontinence. It is essential to have any other pathology or urinary tract infection excluded.

The tibial nerve can be stimulated percutaneously at the ankle joint at the medial malleolus, hence the term tibial nerve stimulation. A large bore needle electrode is inserted about 5 cm cephalad of the medial malleolus and behind the tibia, with an electrode on the surface of the arch of the foot. Initially animal studies confirmed that stimulation at the hindfoot afferents to hind limb muscles resulted in inhibition of bladder activity (McPherson, 1966a). This did not occur when stimulating the cutaneous afferents. This was followed up by further animal and human studies (McGuire et al., 1983a,b; McPherson, 1966b; Sato et al., 1983). It is not clearly understood how stimulation of the posterior tibial nerve at the medial malleolus results in inhibition or reduction of urinary bladder activity. McPherson (1966b) proposed that the pathway was mediated in the forebrain, as intercollicular decerebration or thoracic spinal cord transection abolished the effect. However, there is no definitive explanation as to the benefit that tibial nerve stimulation works in the neural circuitry.

3.3.12.2.4 Tibial Nerve Injury

Injury of the tibial nerve is not common due to its deep location within the popliteal fossa. However, deep trauma, for example, lacerations could potentially result in severance of the nerve. In addition, a severe posterior dislocation of the knee, for example, from a road traffic accident, which may also happen with fractures, may result in tibial nerve injury.

Typically, a patient with a tibial nerve injury will be unable to plantar flex their ankle or toes. Sensation may be absent on the sole of the foot.

3.3.13 Common Peroneal Nerve
The common peroneal nerve (Figure 3.12) arises from the fourth and fifth lumbar vertebrae and the first and second sacral vertebrae (L4–S2). It is typically entwined with the sciatic nerve in the gluteal region and also the thigh as it descends toward the popliteal fossa as its own defined nerve. It is approximately half the size of the tibial nerve. It follows the biceps femoris very closely, and is sometimes concealed by it. The common peroneal nerve then crosses the lateral head of the gastrocnemius superficially. It then winds round the head of the fibula at its posterior aspect, under the cover of the peroneus longus. Roughly at this point, it will divide into its terminal branches, but first gives off cutaneous and articular branches.

Typically, the common peroneal nerve gives off the superficial and deep peroneal nerves, and also an articular branch. The articular branch gathers sensory information from the superior tibiofibular joint and the capsular ligament of the knee at its anterolateral portion, and is a common site for intraneural ganglia (Spinner et al., 2007). There are also two cutaneous branches – the lateral sural cutaneous nerve of the calf and the sural communicating branch. The lateral sural cutaneous nerve supplies sensation to the proximal portion of the leg (anterior, posterior, and lateral). The sural communicating nerve, also referred to as the peroneal communicating branch, arises from approximately 11 mm above the joint line (Deutsch et al., 1999). The other two major branches of the common peroneal nerve are referred to as the superficial and deep peroneal nerves. The summary of the branches, and what they supply is given in Table 3.12.

3.3.13.1 Superficial Peroneal Nerve
This branch initially starts deep to the peroneus longus, at the main bifurcation of the common peroneal nerve. It then courses inferiorly and

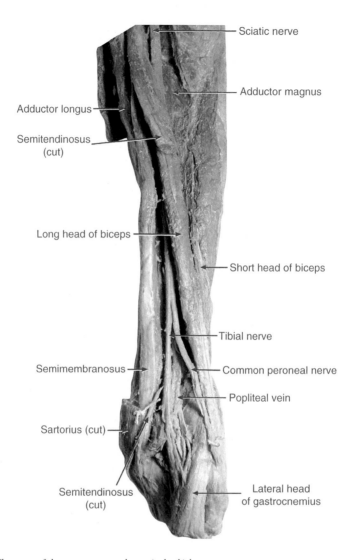

Sciatic nerve

Adductor magnus

Adductor longus

Semitendinosus (cut)

Long head of biceps

Short head of biceps

Tibial nerve

Semimembranosus

Common peroneal nerve

Popliteal vein

Sartorius (cut)

Semitendinosus (cut)

Lateral head of gastrocnemius

Fig. 3.12. The course of the common peroneal nerve in the thigh.

anteriorly between the peronei (peroneus longus and brevis, supplying them as it courses by them), and extensor digitorum longus. The superficial peroneal nerve will then pierce the deep fascia of the leg in its lower one-third, giving off its final branches – the lateral and medial.

The *lateral branch* of the superficial peroneal nerve courses along the dorsum of the foot and supplies the sides of the toes touching between the third and fourth, and the fourth and fifth toes. This branch

Table 3.12 The Main Branches and Subdivisions of Each of the Branches from the Common Peroneal Nerve

Common Peroneal Nerve

Branch		Structures/area supplied
Articular		Tibiofibular joint (proximal), capsular ligament of the knee
Cutaneous	Lateral sural cutaneous	Proximal portion of the leg
	Sural communicating	Joins with sural nerve
Superficial		Peroneus (fibularis) longus and brevis
Lateral branch (cutaneous innervation)		Lateral aspect of third toe, fourth toe, medial aspect of the fifth toe
Medial branch (cutaneous innervation)		Medial side of great toe, lateral side of second toe, medial side of third toe
Deep		Tibialis anterior, extensor digitorum longus, peroneus (fibularis) tertius, extensor hallucis longus
Lateral branch		*Motor*: extensor digitorum brevis, 1st dorsal interosseous muscle; *sensory*: 2nd–4th toes for their tarsal and metatarsophalangeal joints
Medial branch		*Motor*: 1st dorsal interosseous muscle; *sensory*: metatarsophalangeal joint of the great toe, lateral aspect of the great toe and the medial aspect of the second toe
It highlights the cutaneous, motor and sensory innervation from each of these.		

communicates with the sural nerve and provides sensory innervation to the lateral aspect of the ankle.

The *medial branch* then goes to provide cutaneous innervation of the medial side of the great toe, and the lateral side of the second toe and the medial side of the third toe.

Note: The sural nerve supplies the outer aspect of the fifth digit.

In essence, the only areas of the foot not supplied by the superficial peroneal nerve, and its branches, are as follows:

1. *Lateral side of little toe:* supplied by the *sural nerve*
2. *Joining side of first and second toes* (i.e., lateral aspect of great toe and medial side of second toe): supplied by the *deep peroneal nerve*

3.3.13.2 Deep Peroneal Nerve
As with the superficial peroneal nerve, it originates at the bifurcation of the common peroneal nerve and is its terminal branch. It continues

winding around the neck of the fibula under the peroneus longus. It then pierces the anterior intermuscular septum and extensor digitorum longus, passing inferiorly on the intermuscular membrane, under the cover of extensor hallucis longus and extensor digitorum longus. It then appears close to the anterior tibial artery found laterally initially to this vessel, then anterior, and then it is found lateral to this vessel again. Both the anterior tibial artery and the deep peroneal nerve are found deep to the extensor retinaculum. The final branches of the deep peroneal nerve are the lateral and medial branches, like the superficial peroneal nerve.

The lateral branch courses over the tarsus, deep to the extensor digitorum brevis, supplying it as it does so. In addition, the lateral branch supplies the 2nd–4th toes for their tarsal and metatarsophalangeal joints. It also supplies the second dorsal interosseous muscle.

The medial branch of the deep peroneal nerve supplies the lateral aspect of the great toe and the medial aspect of the second toe. It also supplies the metatarsophalangeal joint of the great toe, and also gives off a twig to the first dorsal interosseous muscle.

3.3.13.3 Clinical Applications
3.3.13.3.1 Common Peroneal Nerve Injury
Pathologies of the common peroneal nerve, and its associated branches, can be classified into two categories – traumatic and compressive lesions (Van den Bergh et al., 2013). For those cases that are traumatic in origin, it is due to the superficial nature of the common peroneal nerve as it winds round the head of the fibula, combined with the epineural supporting tissue at this site (Van den Bergh et al., 2013). Of those traumatic cases, they can be due to penetrating causes, bruising or traction, often some or all of these coexisting together (Valle and Zamorani, 2007). In cases where there is compression, again, there can be a variety of etiologies. These typically can be intraneural, for example, ganglia, peripheral nerve sheath tumors, or extraneural compressive lesions, for example, osteochondroma or extraneural ganglia.

3.3.13.4 Clinical Presentation
As the common peroneal (fibular) nerve supplies ALL the muscles in the anterior and lateral compartments of the leg, the resultant

effect of a pathology affecting this nerve is *footdrop*. Therefore, the dorsiflexors of the ankle and also the evertors of the foot will be paralyzed.

The reason for this is that the common peroneal nerve supplies all the dorsiflexors at the ankle and also the peroneal muscles, responsible for eversion of the foot. So, as well as an inability to dorsiflex the foot, there is unopposed inversion of the foot, due to lack of innervation of the evertors (peroneal muscles). In essence, the patient will walk as if the limb on the affected side is *too long*. This means that during the swing phase of the gait cycle, the foot will not clear the ground adequately.

The other features which may also be present in a patient with a common peroneal nerve palsy could be:

1. *Waddling gait*: this is because when the affected limb is raised from the ground, the patient will lean over to the unaffected side to help provide height to raise the foot from the ground.
2. *Swing-out gait*: this is seen as the patient tries to allow a lateral increase in length of the limb to aid the toes clearing the ground during the swing phase of walking.
3. An increased *steppage gait*: this is where there is increased flexion at both the hip and knee joint to help raise the foot from the ground, increasing the vertical height.

All of these signs are, in essence, to aid bringing the foot and toes from the ground during the swing phase of the gait cycle to enable them to walk adequately.

As walking is really seen as controlled falling, during the end of the swing phase of the gait cycle, and to heel strike, the dorsiflexors tend to act as a brake for the lower limb to slow it down. As the heel and foot come to rest during the stance phase, typically dorsiflexors eccentric contraction would slow the foot as it comes to rest on the ground. In those patients with a common peroneal nerve palsy, this contraction of the dorsiflexors does not happen. Therefore, the foot ends up slamming the floor with a classic *clop* noise. This also results in an increase in force transmitted through the foot, ankle and upward to the knee joint via the tibia.

3.3.13.5 Clinical Examination

Examination of the ankle and foot should test for the power of *dorsiflexors of the ankle* (against resistance from the clinician), *foot inversion*, and *eversion* and also sensation should be checked at the ankle and all of the foot. Sensation may be altered or absent on the *dorsum of the foot* and on the *anterolateral aspect of the leg*. The examiner should also ensure analysis of gait is observed to identify any of the features mentioned affecting walking.

3.3.13.5.1 Strawberry Picker's Palsy

Koller and Blank (1980) described an unusual case of a 22-year-old migrant worker in the United States. The patient described in the case presentation was a strawberry picker and worked frequently in a crouched position with *duckwalking* forward movements. He presented with severe yet incomplete footdrop but with normal strength in the gastrocnemius, ankle invertors, and intrinsic foot muscles. The only sign on examination was hypoesthesia over the lateral aspect of the legs bilaterally.

It was described that because of the position of the work during the day, the common peroneal nerve was compressed between the head of the fibula with the biceps tendon and the lateral head of gastrocnemius, because of the body weight pressing downward during the crouching position, as previously described (Staal et al., 1965). This more unusual presentation of a bilateral compression of the common peroneal nerve with bilateral footdrop was referred to as Strawberry picker's palsy.

3.3.14 Superficial Peroneal Nerve

The superficial peroneal nerve descends anterior to the fibula and between the peroneal muscles and the extensor digitorum longus. In the inferior aspect of the leg, it divides into medial and intermediate dorsal cutaneous branches, which go anterior to the extensor retinacula supplying the toes, as previously mentioned.

The muscular branches pass to the peroneus longus and brevis. There is occasionally an accessory deep peroneal nerve where the nerve to peroneus brevis passes further to the lateral malleolus and ends in twigs to the extensor digitorum brevis and nearby joints and ligaments.

The two terminal cutaneous divisions give twigs to the anterior aspect of the lower leg and dorsum of the foot. The medial of the two branches can also be called the medial dorsal cutaneous nerve. This divides into a branch for the medial side of the great toe, which can also communicate with the branch from the deep peroneal nerve, and a branch, which divides into dorsal digital branches for the adjacent sides of the second and third toes. The lateral of the two branches can be referred to as the intermediate dorsal cutaneous nerve, which then divides into two branches. Each of these divide into dorsal digital nerves for the adjacent sides of the third and fourth, and the fourth and fifth toes. However, there is considerable variation in this, as well as overlap between the territories. The plantar digital branches of the medial and lateral plantar nerves supply the nails and tips of toes.

3.3.15 Deep Peroneal Nerve

The deep peroneal nerve continues winding round the neck of the fibula, under the cover of the peroneus longus and descends down through the leg in close relationship to the tibial artery as it descends to its terminal branches as the medial and lateral branches.

The muscular branches are given to the *extensor hallucis longus*, *extensor digitorum longus*, *tibialis anterior*, and *peroneus tertius*. From this, there is also an articular branch, which passes to the *ankle joint*. Of the terminal branches, the medial, or digital branch lies lateral to the *dorsalis pedis* artery and divides into *dorsal digital nerves* for the adjacent sides of the first and second toes. It can also give off articular branches here and can communicate with the superficial peroneal nerve. The lateral branch goes laterally over the tarsus, deeper to the extensor digitorum brevis, which it supplies, as well as providing some local articular branches here. It can also supply the first three dorsal interossei.

REFERENCES

Abitbol, J.J., Gendron, D., Laurin, C.A., Beaulieu, M.A., 1990. Gluteal nerve damage following total hip arthroplasty: a prospective analysis. J. Arthroplasty 5, 319–322.

Ahmad, M., Tsang, K., Mackenney, P.J., Adedapo, A.O., 2012. Tarsal tunnel syndrome: a literature review. J. Foot Ankle Surg. 18, 149–152.

Akita, K., Sakamoto, H., Sato, T., 1992. Stratification relationship among the main nerves from the dorsal division of the sacral plexus and the innervation of the piriformis. Anat. Rec. 233, 633–642.

Akkaya, T., Comert, A., Kendir, S., Acar, H.I., Gumus, H., Tekdemir, I., Elhan, A., 2008. Detailed anatomy of accessory obturator nerve blockade. Minerva Anestesiol. 74, 119–122.

Al-Alami, A.A., Alameddine, M.S., Orompurath, M.J., 2013. New approach of ultrasound-guided genitofemoral nerve block in addition to ilioinguinal/iliohypogastirc nerve block for surgical anesthesia in two high risk patients: case report. Open J. Anesthesiol. 3, 298–300.

Andersen, H.L., Gyrn, J., Møller, L., Christensen, B., Zaric, D., 2013. Continuous saphenous nerve block as supplement to single-dose local infiltration analgesia for postoperative pain management after total knee arthroplasty. Reg. Anesth. Pain Med. 38, 106–111.

Anson, B.J., McVay, C.B., 1971. Surgical Anatomy, fifth ed. Saunders, Philadelphia, pp. 1083–1089.

Aung, H.H., Sakamoto, H., Akita, K., Sato, T., 2001. Anatomical study of the obturator internus, gemelli and quadratus femoris muscles with special reference to their innervation. Anat. Rec. 263, 41–52.

Babinski, M.A., Machado, F.A., Costa, W.S., 2003. A rare variation in the high division of the sciatic nerve surrounding the superior gemellus muscle. Eur. J. Morphol. 41, 41–42.

Baima, J., Krivickas, L., 2008. Evaluation and treatment of peroneal neuropathy. Curr. Rev. Musculoskelet. Med. 1, 147–153.

Bano, A., Karantanas, A., Pasku, D., Datseris, G., Tzanakakis, G., Katonis, P., 2010. Persistent sciatica induced by quadratus femoris muscle tear and treated by surgical decompression: a case report. J. Med. Case Reports 4, 236.

Berthelot, J.M., Delecrin, J., Maugars, Y., Caillon, F., Prost, A., 1996. A potentially underrecognized and treatable cause of chronic back pain: entrapment neuropathy of the cluneal nerves. J. Rheumatol. 23, 2179–2181.

Borges, L.F., Hallett, M., Selkoe, D.J., Welch, K., 1981. The anterior tarsal tunnel syndrome: report of two cases. J. Neurosurg. 54, 89–92.

Bos, J.C., Stoeckart, R., Klooswijk, A.I., van Linge, B., Bahadoer, R., 1994. The surgical anatomy of the superior gluteal nerve and anatomical radiologic bases of the direct lateral approach to the hip. Surg. Radiol. Anat. 16, 253–258.

Bouaziz, H., Vial, F., Jochum, D., Maccalou, D., Heck, M., Meuret, P., Braun, M., Laxenaire, M.-C., 2002. An evaluation of the cutaneous distribution after obturator nerve block. Anesth. Analg. 94, 445–449.

Campos, N.A., Chiles, J.H., Plunkett, A.R., 2009. Ultrasound guided cryoablation for genitofemoral nerve for chronic inguinal pain. Pain Physician. 12, 997–1000.

Carai, A., Fenu, G., Sechi, E., Crotti, F.M., Montella, A., 2009. Anatomical variability of the lateral femoral cutaneous nerve: findings from a surgical series. Clin. Anat. 22, 365–370.

Central Manchester University Hospitals. http://www.cmft.nhs.uk/directorates/mentor/documents/InjectionTechnique.pdf (accessed 23.02.2015.).

Choquet, O., Capdevila, X., Bennourine, K., Feugeas, J.L., Bringuier-Branchereau, S., Manelli, J.C., 2005. A new inguinal approach for the obturator nerve block: anatomical and randomized clinical studies. Anesthesiology 103, 1238–1245.

Chou, D., Storm, P.B., Campbell, J.N., 2004. Vulnerability of the subcostal nerve to injury during bone graft harvesting from the iliac crest. J. Neurosurg. Spine 1, 87–89.

Christos, S.C., Chiampas, G., Offman, R., Rifenburg, R., 2010. Ultrasound-guided three-in-one nerve block for femur fractures. West J. Emerg. Med. 11 (4), 310–313.

Colterjohn, N.R., Bednar, D.A., 1997. Procurement of bone graft from the iliac crest. An operative approach with decreased morbidity. J. Bone Joint Surg. Am. 79, 756–759.

Comstock, C., Imrie, S., Goodman, S.B., 1994. A clinical and radiographic study of the "safe area" using the direct lateral approach for total hip arthroplasty. J. Arthroplasty 9, 527–531.

Delawi, D., Dhert, W.J., Castelein, R.M., Verbout, A.J., Oner, F.C., 2007. The incidence of donor site pain after bone graft harvesting from the posterior iliac crest may be overestimated: a study on spine fracture patients. Spine 32, 1865–1868.

Deutsch, A., Wyzykowski, R.J., Victoroff, B.N., 1999. Evaluation of the anatomy of the common peroneal nerve. Defining nerve-at-risk in arthroscopically assisted lateral meniscus repair. Am. J. Sports Med. 27 (1), 10–15.

Gulec, A., Buyukbebeci, A.O., 1996. Late superior gluteal nerve palsy following posterior fracture-dislocation of the hip. Acta Orthop. Belg. 62, 218–221.

Frymoyer, J.W., Howe, J., Kuhlmann, D., 1978. The long-term effects of spinal fusion on the sacro-iliac joints and ilium. Clin. Orthop. Relat. Res. 134, 196–201.

Hallgren, A., Björkman, A., Chemnitz, A., Dahlin, L.B., 2013. Subjective outcome related to donor site morbidity after sural nerve graft harvesting: a survey in 41 patients. BCM Surgery 13, 39.

Hammer, W.I., 2007. Functional soft-tissue examination and treatment by manual methods, third ed. Jones and Bartlett Publishers Inc., MA, USA.

Havel, P.E., Ebraheim, N.A., Clark, S.E., 1988. Tibial branching in the tarsal tunnel. Foot Ankle 9, 117–119.

Hodges, P.W., McLean, L., Hodder, J., 2014. Insight into the function of the obturator internus muscle in humans: observations with development and validation of an electromyography recording technique. J. Electromyogr. Kinesiol. 24, 489–496.

Hospodar, P.P., Asman, E.S., Traub, J.A., 1999. Anatomic study of the femoral cutaneous nerve with respect to the ilioinguinal surgical dissection. J. Orthop. Trauma. 13, 17–19.

International Standards for Neurological Classification of Spinal Cord Injury (ISNCSCI). American Spinal Injury Association. http://www.asia-spinalinjury.org/elearning/ASIA_ISCOS_high.pdf (accessed 13.03.2015.).

Jacobs, L.G., Buxton, R.A., 1989. The course of the superior gluteal nerve in the lateral approach to the hip. J. Bone Joint Surg. Am. 71, 1239–1243.

Kassarijian, A., 2008. Signal abnormalities in the quadratus femoris muscle: tear or impingement. Am. J. Roentgenol. 190, 380–381.

Khan, U.A., Krishnamoorthy, B., Najam, O., Waterworth, P., Fildes, J.E., Yonan, N., 2010. A comparative analysis of saphenous vein conduit harvesting technqiues for coronary artery by-pass grafting – standard bridging versus open technique. Interact. Cardiovasc. Thorac. Surg. 10, 27–31.

Koller, R.L., Blank, N.K., 1980. Strawberry Picker's Palsy. Arch. Neurol. 37, 320.

Kuniya, H., Aota, Y., Saito, T., Kamiya, Y., Funakoshi, K., Terayama, H., Itoh, M., 2013. Anatomical study of superior cluneal nerve entrapment. J. Neurosurg. Spine 19, 76–80.

Lang, S.A., Yip, R.W., Chang, P.C., Gerard, M.A., 1993. The femoral 3-in-1 block revisited. J. Clin. Anesth. 5, 292–296.

Lavigne, P., Loriot de Rouvray, T.H., 1994. The superior gluteal nerve. Anatomical study of its extrapelvic portion and surgical resolution by transgluteal approach. Rev. Chir. Orthop. Reparatrice Appar. Mot. 80, 188–195.

Lindenbaum, B.L., 1979. Ski boot compression syndrome. Clin. Orthop. Relat. Res., 109–110.

Ling, Z.X., Kumar, V.P., 2006. The course of the inferior gluteal nerve in the posterior approach to the hip. Bone Joint J. 88 (B), 1580–1583.

Lotero, M.A.A., Diaz, R.R., Aguilar, M.A.M., Jaramillo, S.J., 2014. Efficacy and safety of ultrasound-guided saphenous nerve block in patients with chronic knee pain. Rev. Colomb. Anestesiol. 42, 166–171.

Macalou, D., Trueck, S., Meuret, P., Heck, M., Vial, F., Ouologuem, S., Capdevila, X., Virion, J.-M., Bouaziz, H., 2004. Postoperative analgesia after total knee replacement: the effect of an obturator nerve block added to the femoral 3-in-1 nerve block. Anesth. Analg. 99, 251–254.

Machado, F.A., Babinski, M.A., Brasil, F.B., Favorito, L.A., Abidu- Figueiredo, M., Costa, M.G., 2003. Incidencia de variaciones anatómicas entre el nervio isquiático y musculo piriformis durante el período fetal humano:10 y 37 semanas póst- concepción. Int. J. Morphol. 21, 29–35.

Magee, D.J., Zachazewski, J.E., Quillen, W.S., 2009. Pathology and Intervention in Musculoskeletal Rehabilitation. Saunders Elsevier, Missouri, USA, p. 679.

Maigne, R., 1980. Low back pain of thoracolumbar origin. Arch. Phys. Mcd. Rehabil. 61, 389–395.

Maigne, J.Y., Lazareth, J.P., Guerin Surville, H., Maigne, R., 1989. The lateral cutaneous branches of the dorsal rami of the thoraco-lumbar junction. an anatomical study on 37 dissections. Surg. Radiol. Anat. 11, 289–293.

Marinacci, A.A., 1968. Medial and anterior tarsal tunnel syndrome. Electromyography 8, 123–134.

Markhede, G., Stener, G., 1981. Function after removal of various hip and thigh muscles for extirpation of tumors. Acta Orthop. Scand. 52, 373.

McGuire, E., Morrissey, S., Zhang, S., Horwinski, E., 1983a. Control of reflex detrusor activity in normal and spinal injured nonhuman primates. J. Urol. 129 (1), 197–199.

McGuire, E.J., Zhang, S.C., Horwinski, E.R., Lytton, B., 1983b. Treatment of motor and sensory detrusor instability by electrical stimulation. J. Urol. 129 (1), 78–79.

McPherson, A., 1966a. The effects of somatic stimuli on the bladder in the cat. J. Physiol. 185, 185–196.

McPherson, A., 1966b. Vesico-somatic reflexes in the chronic spinal cat. J. Physiol. 185, 197–204.

Melendez, M.M., Glickman, L.T., Dellon, A.L., 2013. Peroneal nerve compression in figure skaters. Clin. Res. Foot Ankle. 1, 1–3.

Mercer, D., Morrell, N.T., Fitzpatrick, J., Silva, S., Child, Z., Miller, R., DeCoster, T.A., 2011. The course of the distal saphenous nerve: a cadaveric investigation and clinical implications. Iowa Orthop. J. 31, 231–235.

Miller, K.J., 2007. Physical assessment of lower extremity radiculopathy and sciatica. J. Chriopr. Med. 6, 75–82.

Moore, W., 2005. The knife man. Blood, body-snatching and the birth of modern surgery. Bantam Books, London, p. 18.

Moore, K.L., Dalley, A.F., 2006. Clinically oriented anatomy, fifth ed. Lippincott Williams and Wilkins, Baltimore, USA, p. 502.

Moucharafieh, R., Wehbe, J., Maalouf, G., 2008. Meralgia paresthetica: a result of tight new trendy low cut trousers ('taille basse'). Int. J. Surg. 6, 164–168.

Nagaoka, M., Matsuzaki, H., 2005. Ultrasonography in tarsal tunnel syndrome. J. Ultrasound. Med. 24, 1035–1040.

NHS Sciatic – Treatment, 2015. http://www.nhs.uk/Conditions/Sciatica/Pages/Treatment.aspx accessed 05.03.2015.].

O'Brien, S.D., Bui-Mansfield, L.T., 2007. MRI of quadratus femoris muscle tear: another cause of hip pain. Am. J. Roentgenol. 189, 1185–1189.

Ogbemudia, A., Bafor, A., Igbinovia, E., Ogbemudia, P.E., 2010. Hip hemiarthroplasty for femoral neck fractures using the modified Stracathroc approach – short term results in twenty-six patients. J. Surg. Tech. Case Rep. 2, 8–12.

Ortigüela, M.E., Wood, M.B., Cahill, D.R., 1987. Anatomy of the sural nerve complex. J. Hand Surg. 12, 1119–1123.

Paraskevas, G., Tzika, M., Natsis, K., 2014. Entrapment of the superficial peroneal nerve: an anatomical insight. J. Am. Podiatr. Med. Assoc., 105(2):150113115057001. DOI: 10.7547/12-151.1.

Patijn, J., Mekhail, N., Hayek, S., Lataster, A., van Kleef, M., Van Zundert, J., 2011. Meralgia Paresthetica. Pain Pract. 11, 302–308.

Pearce, J.M.S., 2006. Meralgia paraesthetica (Bernhardt-Roth syndrome). J. Neurol. Neurosurg. Psychiatry 77, 84.

Peng, P.W., Tumber, P.S., 2008. Ultrasound-guided interventional procedures for patients with chronic pelvic pain – a description of techniques and review of literature. Pain Physician. 11, 215–224.

Rea, P., 2015. Essential Clinical Anatomy of the Nervous System, first ed. Elsevier Academic Press, San Diego, ISBN 9780128020302.

Reade, B.M., Longo, D.C., Keller, M.C., 2001. Tarsal tunnel syndrome. Clin. Podiatr. Med. Surg. 18, 395–408.

Russell, S.M., 2006. Examination of Peripheral Nerve Injuries – An Anatomical Approach. Thieme Medical Publishers Inc, NY, USA, p. 131.

Sato, A., Sato, Y., Schmidt, R.F., Torigata, Y., 1983. Somato-vesical reflexes in chronic spinal cats. J. Auton. Nerv. Syst. 7, 351–362.

Smoll, N.R., 2010. Variations of the piriformis and sciatic nerve with clinical consequence: a review. Clin. Anat. 23, 8–17.

Soong, J., Schafhalter-Zoppoth, I., Gray, A.T., 2007. Sonographic imaging of the obturator nerve for regional block. Region. Anesth. Pain Med. 32, 146–151.

Spinner, R.J., Desy, N.M., Rock, M.G., Amrami, K.K., 2007. Peroneal intraneural ganglia. Part I. Techniques for successful diagnosis and treatment. Neurosurg. Focus. 22 (6), E16.

Staal, A., deWeerdt, C.J., Went, L.N., 1965. Hereditary compression syndrome of peripheral nerves. Neurology 15, 1008–1017.

Stewart, W.F., Van Rooyen, J.B., Cundiff, G.W., Abrams, P., Herzog, A.R., Corey, R., Hunt, T.L., Wein, A.J., 2003. Prevalence and burden of overactive bladder in the United States. World J. Urol. 20, 327–336.

Strong, E.K., Davila, J.C., 1957. The cluneal nerve syndrome; a distinct type of low back pain. Ind. Med. Surg. 26, 417–429.

Talu, G.K., Ozyalçin, S., Talu, U., 2000. Superior cluneal nerve entrapment. Reg. Anesth. Pain Med. 25, 648–650.

Thallaj, A., Rabah, D., 2011. Efficacy of ultrasound-guided obturator nerve block in transuretheral surgery. Saudi J. Anaesth. 5, 42–44.

Tillmann, B., 1979. Variations in the pathway of the inferior gluteal nerve. Anat. Anz. 145, 293–302.

Topçu, I., Aysel, I., 2014. Ultrasound guided posterior femoral cutaneous nerve block. AĞRI 26 (3), 145–148.

Trendelenburg, F., 1895. Ueber den Gang bei angeborener Huftgelenksluxation. Dtsch Med Wochenschr 21, 21.

Valle, M., Zamorani, M.P., 2007. Nerve and blood vessels. In: Bianchi, S., Martinoli, C. (Eds.), Ultrasound of the Musculoskeletal System. Springer, Berlin, pp. 97–136.

Van den Bergh, F.R.A., Vanhoenacker, F.M., De Smet, E., Huysse, W., Verstraete, K.L., 2013. Peroneal nerve: normal anatomy and pathologic findings on routine MRI of the knee. Insights Imaging 4, 287–299.

van Slobbe, A.M., Bohnen, A.M., Bernsen, R.M., Koes, B.W., Bierma-Zeinstra, S.M., 2004. Incidence rates and determinants in meralgia paresthetica in general practice. J. Neurol. 251, 294–297.

Van Ramshorst, G.H., Kleinrensink, G.-J., Hermans, J.J., Terkivatan, T., Lange, J.F., 2009. Abdominal wall paresis as a complication of laparoscopic surgery. Hernia. 13, 539–543.

Vloka, J.D., Hadžić, A., Ernest, A., Daniel, T., 2001. The division of the sciatic nerve in the popliteal fossa: anatomical implications for popliteal blockade. Anesth. Analg. 92, 215–217.

White, B., Gonzalez, P., Malanga, G.A., Akuthota, V., 2009. Physical examination of the peripheral nerves and vasculature. In: Akuthota, V., Herring, S.A. (Eds.), Nerve and Vascular Injuries in Sports Medicine. Springer Science+Business Media, New York, p. 49.

Williams, P., Warwick, R., Dyson, M., Bannister, L.H., 1989. Gray's anatomy, thirty-seventh ed. Churchill Livingstone, London, p. 599.

Yarwood, J., Berrill, A., 2010. Nerve blocks of the anterior abdominal wall. Contin. Educ. Anaesth. Crti. Care Pain. 10, 182–186.

Zenati, M.A., Shroyer, A.L., Collins, J.F., Hattler, B., Ota, T., Almassi, G.H., Amidi, M., Novitzky, D., Grover, F.L., Sonel, A.F., May 1–5, 2010. Impact of endoscopic versus open saphenous vein harvest technique on late coronary artery bypass grafting patient outcomes in the ROOBY (Randomized On/Off Bypass) Trial. 90th Annual Meeting of the American Association for Thoracic Surgery. Toronto, Ontario, Canada.

Zhu, J., Zhao, Y., Liu, F., Huang, Y., Shao, J., Hu, B., 2012. Ultrasound of the lateral femoral cutaneous nerve in asymptomatic adults. BMC Musculoskelet. Disord. 13, 227.

SUBJECT INDEX

A

Abdomen, laproscopic examination of, 103
Abdominal musculature, 104
    external oblique, 104
    internal oblique, 104
Abductor hallucis, 119
Accessory nerve
    adductor brevis, 154
    adductor longus, 152
    adductor magnus, 152
    gracilis, 154
    obturator externus, 154
    pectineus, 154
Accessory phrenic nerve, 57
Acromioclavicular joint, 51, 53
Adductor canal, 141
Adductor muscle strength, 111
Afferent fibers, 15
    general somatic, 15
    general visceral, 15
    special visceral, 16
American Spinal Injury Association
    (ASIA), 17, 106, 107
Analgesia
    intraoperative, 104
    postoperative, 104
Anaxonic neurons, 4
Ankle jerk reflex, 162
Antebrachial cutaneous nerve, lateral,
    64, 65
Anterior intermuscular septum, 116
Aponeurosis, 105
Arc syndrome test, 56
ASIA. *See* American Spinal Injury
    Association (ASIA)
Astrocytes, 3
    metabolite exchange, 3
    structural integrity, 3
Autonomic ganglion, 18
Axillary brachial plexus block, 90
Axillary nerve, 88
    anterior branch of, 88
    clinical applications, 90
    clinical examination, 89
    innervated muscles, 88
    posterior branch of, 88
    sensory assessment, 89
Axons, 1, 3, 5, 9–11, 13
    betz neurons, 4
    multipolar neurons, 4
    myelinated, 5

B

Babinski sign, 34, 37
Bernhardt–Roth syndrome, 109
Betz neurons, 4
Biceps femoris, 146
Bicipital aponeurosis, 67, 68
Bipedal locomotion, 101
Bipolar neurons, 4
Bone grafting, 121
Bony pelvis, 101
Brachial cutaneous nerve, medial, 75
Brachialis fascia, 67
Brachial plexus, 90
    anatomical location of, 44
    branches of, 41
    cords of, 43
    dorsal scapular nerve, 45
    lateral cord, 43
    lower trunk, 43
    medial cord, 43
    middle trunk, 43
    and omohyoid muscle, 43
    posterior cord, 43
    posterior triangle, 42
    postfixed, 42
    prefixed, 42
    and sternocleidomastoid
        muscle, 43
    surface anatomy, 43
    ventral rami of, 43
Brachioradialis, 92
Brachydactyly, 62
Bragaard's test, 149
Brain, 6
    forebrain, 6
    herniation, 7
    hindbrain, 6
    midbrain, 6

Branchial arches, 23
Bulbospongiosus, 136

C
CABG. *See* Coronary artery bypass
    surgery (CABG)
CAD. *See* Coronary artery disease
    (CAD)
Cardiovascular system (CVS), 28
Carpal tunnel, 69
  anatomical variants within, 70
  flexor digitorum
    profundus, 70
    superficialis, 70
  flexor pollicis longus, 70
  median artery, 70
  median nerve, 70
  syndrome, 70
    causes of, 75
    investigation of, 75
    Phalen's sign, 75
    Tinel's test, 74
    treatment of, 75
  tendons, 70
Cell body, 1, 4
Central nervous system (CNS), 1, 2. *See
    also* Nervous system
  brain, 6
  forebrain, 7
  gray and white matter, 5
  hindbrain, 9
  hypothalamus, 8
  laminae, 2
  midbrain, 9
  neuroglia, 2
  neurons, 1
  spinal cord, 9
  thalamus, 8
Cerebellar disease, 38
Cerebral aqueduct, 6, 9
Cerebrum, 6
  basal ganglia, 8
  cerebral cortex, 7
    frontal lobes, 8
    occipital lobes, 8
    parietal lobes, 8
    temporal lobes, 8
  limbic system, 8

Cervical vertebral nerves, 45, 57
Clunial nerves, 119
  clinical applications, 121
    entrapment neuropathy, 121
    injury to, 121
  inferior, 120
  medial, 120
  superior, 119
  syndrome, 121
Clunial nerves, inferior, 112
Collateral ganglia, 19
Connexons, 2
Coracobrachialis, 63
Coronary artery bypass surgery (CABG),
    114
Coronary artery disease (CAD), 114
Corpus cavernosum penis, 137
Cranial nerves, 6, 15, 16, 20, 21, 39
  accessory, 15
  general visceral afferent, 15
  motor, 22
  parasympathetic nervous system, 20
  peripheral nervous system, 15
  twelve pairs of, 24
Cubital tunnel, 82
  syndrome, 83
Cutaneous nerve, medial
  of arm, testing of, 76
  of forearm, 76, 77
    clinical examination, 76, 77
    testing of, 78
Cutaneous nerve of thigh, lateral, 108
  anterior cutaneous branches, 109
  clinical applications, 109
CVS. *See* Cardiovascular system
    (CVS)

D
Deep fibular nerve, 116
Deep peroneal nerve, 168
  clinical applications, 169
  clinical examination, 171
  clinical presentation, 169
Deltoid muscle
  assessment of, 89
  intramuscular injection to, 90
Dendrites, 1, 4, 5, 10, 11
Dermatome, 16, 33, 78, 102, 105–108

Dermatome maps, 102
  axial line, 102
Diencephalon, 6, 7
Dorsal cutaneous nerve, intermediate, 115
Dorsal cutaneous nerve, medial, 115
Dorsal nerves, 93, 137
  of clitoris, 137
  innervation territories of, 93
  of penis, 137
Dorsal scapular nerve, 45, 46
  clinical applications, 47
  clinical examination, 45
  entrapment, 48
  levator scapulae, 45
  rhomboid major muscles, 45
  scalenus medius, 45
  syndrome, 47
    abnormal shoulder joing movement, 47
    treatment of, 48
    winged scapula, 47
Dysdiadochokinesis, 38

E
Edinger–Westphal nucleus, 20
Efferent fibers, 15
  general visceral, 15
  somatic, 15
  special visceral, 15
Electromyography, 129
Embryological development, 22
Epicondylitis examination, 67
Erb–Duchenne palsy, 60
Extensor digitorum longus, 114, 116

F
Femoral cutaneous nerve
  intermediate, 142
  lateral, 108
  medial, 142
  posterior, 111
Femoral nerve, 139, 142
  branches, 112
  clinical applications, 143
  clinical examination, 144
Femoral triangle, 141
Fibular nerve, superficial, 114

clinical applications, 115
clinical examination, 115
Filum terminale, 10
Finger–nose test, 38
Flexor carpi
  radialis, 68, 70
  ulnaris, 70, 83
Flexor digitorum
  brevis, 119
  superficialis, 68, 73, 92
Flexor pollicis longus, 92
Flexor retinaculum, 68, 69
Forebrain, 6–8, 21, 165

G
Ganglia, 2, 5, 14, 19, 169
  extraneural, 169
  intraneural, 166
Gastrointestinal system (GI), 18, 29
Genitofemoral nerve, 105, 143
  clinical applications, 105
  light touch examination, 106
  sensory assessment, classification of, 108
Genitourinary system, 31
Glenohumeral joint, 53, 55
Gluteal nerve, inferior, 127
  clinical applications, 129
  clinical examination, 128
Gluteal nerve innervations, 129
Gluteal skyline test, 128
Gluteus
  maximus, 112, 127
    gluteofemoral, 128
    ischial bursae, 128
    trochanteric, 128
  medius, 123, 124
  minimus, 124
Gray and white matter, 5
  brain communication, 5
  glial cells, 5
  myelinated axons, 5
Great saphenous vein, 113
Guyon's canal, 72
Guyon's tunnel syndrome, 83

H
Hindbrain, 6, 9, 22
Hip joint, 101

Hip joint *(cont.)*
    congenital dislocation of, 133
    erect examination, 130
    examination of, 130
    supine patient examination, 131
        hip stability, 133
        inspection, 131
        leg length measurement, 132
        movement, 132
        palpation, 132
Hip replacements, 129
Hornblower's test, 89
Hunter's canal, 141
Hypothalamus, 8

I
Iliac crest, 103
Iliac spine, 103, 109, 110
Iliohypogastric nerves, 103, 104
Infraspinatus, 51, 52
Infraspinous fossa, 52
International Standards for Neurological
    Classification of Spinal Cord Injury
    (ISNCSCI), 17, 33, 35, 73, 76, 78, 106
Ischial tuberosity, 134
Ischiocavernosus, 136
ISNCSCI. *See* International Standards
    for Neurological Classification of
    Spinal Cord Injury (ISNCSCI)

L
Labia majora, 137
Laborer's nerve, 79
Lasègue's sign, 149
Latissimus dorsi, 85, 86
    as adductor, 86
    as extensor, 86
    spinous processes, 86
Levator ani, 136
Levator scapulae, 45
    muscles, power of, 47
    palpate, 47
Light touch sensation, 107
Limb nervous system, lower, 101.
        See also Nervous system
    bones, existing in, 101, 102
    development of, 101
    parts of, 101

Linburg–Comstock syndrome, 72
Lipoproteins, 1
Lower limb, cutaneous innervation of, 102
Lumbar puncture, 10
Lumbar vertebrae, 102
Lumbosacral plexus, 102

M
Macroglia, 3
    astrocytes, 3
        metabolite exchange, 3
        structural integrity, 3
    enteric glia
        digestion, 3
        homeostasis, maintenance of, 3
    ependymal cells, 3
        choroidal, 3
        ependymocytes, 3
        tanycytes, 3
    oligodendrocytes, 3
    radial glia, 3
    satellite cells, 3
    schwann cells
        phagocytosis, 3
Maigne's syndrome, 122
Marinacci communication, 72
Martin-Gruber anastomosis, 72
Median cubital vein, 76
Median nerve, 67
    branches of, 69
    carpal tunnel, 69
    clinical applications, 74
    clinical examination, 73, 74
    location of, 68
    motor branch, variation in, 72
    muscles innervated, 71
    muscular variants, 72
    proximal bifurcation of, 70
    structures, innervation of, 69
Meningitis, 10
Meralgia paraesthetica, 109
Midbrain, 9
    cerebral aqueduct, 9
    cerebral peduncles, 9
    tegmentum, 9
Motor system
    coordination, 38
    examination of, 34

inspection, 34
motor and sensory nerves, 122
power, 35
superior gluteal nerve, 122
clinical applications, 125
inferior branch, 123
superior branch, 123
tendon reflexes, 36
lower limbs, 37
upper limbs, 36
tone, 34
Multipolar neurons, 4
Musculocutaneous nerve, 63, 64
bicipital aponeurosis, 67
clinical applications, 67
clinical examination, 65
inspection, 66
movement, 66
palpation, 66
Musculoskeletal system, 32
Musician's nerve, 79
Myelin sheath, 1, 3
Myotome, 17

N
Nerve sensation, testing of, 111
Nerve to subclavius, 57
clinical applications, 58
clinical examination, 57
erb's point, 60
inspection, 58
palpation, 58
Nervous system. *See also* Central nervous
system (CNS)
autonomic, 15, 18
divisions of, 1
peripheral nervous system, 14
spinal nerves, 16
family history, 28
functional division of, 17
autonomic nervous system, 18
parasympathetic nervous system, 20
somatic nervous system, 17
sympathetic nervous system, 19
upper limb, 41
Neuroglia, 2
macroglia, 2
microglia, 2

Neurological system, 31
Neurons, 1
anaxonic, 4
betz, 4
bipolar, 4
interneurons, 1
basket cells, 5
lugaro cells, 5
spindle cells, 5
unipolar brush cells, 5
motor, 1
multipolar, 4
pseudounipolar, 4
sensory, 1
unipolar, 4
Neurotransmitter, 2, 17, 21
cholinergic, 21

O
OAB. *See* Overactive bladder (OAB)
Obturator internus, 137, 139
Obturator nerve, 151, 153
accessory nerve, 152
anterior branch, 152
clinical applications, 154
cutaneous nerve of, 110
adductor longus, 110
brevis, 110
clinical applications, 111
clinical examination, 111
gracilis, 110
pectineus, 110
posterior branch, 152
Olfactory epithelium, 4
Omohyoid muscle, 43
Ortolani's test, 133
Overactive bladder (OAB), 165

P
Paget–Schroetter syndrome, 60
Palmaris longus, 68, 70
variability in, 72
Parasympathetic nervous system, 20.
See also Nervous system
cranial nerves, 21
Patient medical history, 23
CVS history, recording of, 28
past history, 27

Patient medical history *(cont.)*
   obstetric, 27
   psychiatric, 28
   surgical, 27
  social history, 28
   alcohol consumed, 28
   occupation, 28
   relationship status, 28
   smoking, 28
Pectineus muscles, 110
Pectoralis major, 61
  examination of, 62
  injury of, 62
Pectoralis minor, 61
Pectoral nerve
  lateral, 61
  medial, 61
   clinical applications, 62
   clinical examination, 62
Pelvic girdle, 101
Perineal nerves, 112, 136
Peripheral nervous system, 14, 44. *See also* Nervous system
  cranial, 14
  spinal, 14
Peroneal nerve
  common, 166–168
  superficial, 114, 166, 171
Phagocytic cells, 2
Phalen's sign, 75
Piriformis, 137
Plantar nerve
  lateral, 119
  medial, 119
Poland syndrome, 62
Posterior cutaneous nerve, 111
  clinical applications, 112
  clinical examination, 112
Posterior triangle, 42
  anterior boundary, 42
  apex of, 43
  base of, 43
  posterior boundary, 42
  roof of, 43
Post ganglionic sympathetic fibers, 135
Preganglionic fiber, 18–21
Presenting complaint, history of, 27
Pronator teres, 68

Psoas major muscle, 109
Pudendal nerve, 135, 136, 138
  inferior rectal nerve, 136
  penis, dorsal nerve of, 136
  perineal nerve, 136
Pyramidalis, 103

Q

Quadratus femoris muscle, 134
  clinical applications, 134
  clinical examination, 134
  hip adduction, 134
  hip joint, external rotation of, 134
Quadratus lumborum, 103
Quadriceps muscles, 110

R

Radial nerve, 90
  arcade of Frohse, 98
  articular branches, 92
  branches of, 95, 96
  clinical applications, 95
  clinical examination, 94
  cutaneous branches, 91
  epicondyle, 91
  humerus shaft, fracture of, 95
  injury at elbow, 98
  injury at wrist and hand, 98
  muscular branches, 92
  nerve roots, 98
  superficial branch, 92
  surface anatomy of, 91
  terminal branch of, 93
  wrist-drop, 97
Ranvier, nodes of, 1
Rectal nerve, inferior, 137
Rectus abdominis, 103
Respiratory system, 29
Rhomboids, 46
  muscles, power of, 47
  palpate, 47
Romberg's test, 38, 39
Rotator cuff muscles, 54
  Hawkins sign, 55
  injury of, 56
   tendinitis, 56
   treatment of, 56
  tear of, 56
  testing, 55

S

Sacrum, 101
Saphenous nerve, 112
  branches of, 113
  clinical applications, 113
    great saphenous vein, harvesting of,
      114
    nerve block, 114
    vein cutdown, 113
SBT. *See* Standard bridging technique
  (SBT)
Sciatic nerve, 144, 145
  articular, 146
  branching of, 146
  clinical applications, 150
  clinical examination, 148
    bowstring sign, 149
    flip test, 149
    Neri's bowing sign, 149
    sitting test, 149
    supine examination of, 148
  muscular, 146
Sensory system
  assessment, 108
    scoring system for, 108
  ataxia, test for, 38
  examination of, 32
    gait, 32
    patient observation, 32
    speech, 32
  nerves, 104
Serratus anterior muscle, 48, 49, 51
Serratus wall test, 50
Shoulder girdle, 41
Slipped disc, 106
Slump knee bend test, 155
SOC. *See* Space-occupying lesion (SOC)
Somatic nervous system, 17. *See also*
  Nervous system
  cell bodies, 17
  motor neurons, 17
  somites, 22
Space-occupying lesion (SOC), 7
Spinal cord, 1, 2, 4, 7, 9, 10, 13, 14, 17,
  19, 34, 106, 137
  cauda equina, 10
Spinal nerves, 9, 10, 15–17, 21, 51, 102,
  105
Spinoglenoid fossa, 51

Standard bridging technique (SBT), 114
Sternoclavicular joint, 53
Sternocleidomastoid muscle, 43
Subclavius muscle, 57
Subclavius posticus, 58
Subcostal nerve, 103, 104
  clinical applications of, 103
    anterior abdominal wall nerve
      blocks, 104
    autologous bone harvesting, 103
    laparoscopic surgery, risk from, 103
Subsartorial canal, 141
Subscapular fossa, 84
Subscapularis muscle, 84
  tears of, 84
Subscapular nerve
  lower, 84
  upper, 84
Substantia gelatinosa, 11, 12
Superficial transverse perineal muscle,
  136
Superior gluteal nerve injury, 126
  buttocks, intramuscular injection in,
    126
  hip fractures, repair of, 126
  hip joint, dislocation of, 126
Suprascapular nerve, 51
  clinical applications, 56
  clinical examination, 53
  inspection, 54
  movement, 54
  palpation, 54
Supraspinatus, 51
Sural nerves, 112, 117
  clinical applications, 118
  lateral sural cutaneous nerve, 118
  medial sural cutaneous nerve, 118
  peroneal communicating branch, 118
Sympathetic nervous system, 19–21
Synapse, 2
  chemical, 2
  electrical, 2
Syndactyly, 62

T

Tarsal tunnel syndrome, 164
Tectum, 6
Tegmentum, 6, 9
Telencephalon, 6

Tendinitis, 56
Tensor fasciae latae, 125
Teres major, 85
Teres minor, 51, 52, 54, 88, 89
Thalamus, 6, 8
Thomas's test, 132
Thoracic nerve, long, 48, 103
    clinical applications, 50
    clinical examination, 49
    external respiratory nerve of Bell, 49
    serratus anterior, 48
    serratus wall test, 50
Thoracic outlet syndrome, 50, 59
    congenital, 59
    trauma, 59
    treatment of, 60
    tumors within neck, 59
Thoracoabdominal nerves, 104
Thoracodorsal nerve, 85
    clinical applications, 87
    clinical examination, 86
Thoracoscapular mechanism, 53, 54
Tibial nerve, 119, 156–158
    branching of, 156, 160
    clinical applications of, 160
    clinical examination, 162
    injury, 165
    stimulation, 165
Tinel's sign, 164
Tinel's test, 74
Transversus abdominis, 103, 104
Traumatic brain injury, 7
Trendelenburg sign, 126
Trocar placement, 103

Tuso nerve, 93
Twelfth thoracic vertebra (T12), 104
    anterior cutaneous branch, 104
    lateral cutaneous branch, 104

U
Ulnar nerve, 78
    branches of, 80
    claw hand, 82
    clinical applications, 82
    clinical examination, 80
    collateral nerve, 92
    cubital tunnel, 82
    deep branch, 79
    dorsal branch, 78
    injury, 82
    muscular innervation of, 81
    palmar branch, 79
    superficial branch, 79
    variation of, 72
    vertebral levels of, 82
Unipolar neurons, 4
    dorsal root ganglia, 4
    pseudounipolar neurons, 4
Upper limbs
    ball and socket joint, 41
    bones within, 42
    movements, 41
    segments of, 41

V
Ventral ramus, 103
Vertebral column, 101
Von Frey fibers, 107

Printed in the United States
By Bookmasters